POWER AND ENERGY SERIES 905

Power System Protection

Volume 2: Systems and methods

Other volumes in this series:

Power System Protection

Volume 2: Systems and methods

Edited by The Electricity Training Association

The Institution of Engineering and Technology

Published by The Institution of Engineering and Technology, London, United Kingdom

First edition © 1969 The Electricity Council
Revised edition © 1981 The Electricity Council
Second revised edition © 1995 Electricity Association Services Ltd

First published 1969 by Macdonald & Co. (Publishers) Ltd
Revised edition 1981 by Peter Peregrinus Ltd on behalf of the IEE
Reprinted with minor corrections 1986
Reprinted 1990
Second revised edition 1995 by the Institution of Electrical Engineers
Reprinted 2006, 2007 by the Institution of Engineering and Technology

The Institution of Engineering and Technology
Michael Faraday House
Six Hills Way, Stevenage
Herts, SG1 2AY, United Kingdom

www.theiet.org

British Library Cataloguing in Publication Data
A catalogue record for this product is available from the British Library

ISBN (10 digit) 0 85296 836 1
ISBN (13 digit) 978-0-85296-836-9

Printed in the UK by University Press, Cambridge
Reprinted in the UK by Lightning Source UK Ltd, Milton Keynes

Contents

Contents

Foreword

The four volumes which make up this publication owe their origin to a correspondence tuition course launched in 1966 by the UK electricity supply industry, written by expert engineers from both the supply industry and manufacturers, and administered by the Electricity Council. The correspondence course continues to be provided to meet the needs of staff in the electricity supply industry throughout the world. Since privatisation of the industry in the UK the course is now provided by the Electricity Training Association, the industry's training organisation.

It became apparent soon after its inception that the work met a widespread need in the UK and overseas for a standard text on a specialised subject. Accordingly, the first edition of Power System Protection was published in book form in 1969 and has since come to be recognised as a comprehensive and valuable guide to concepts, practices and equipment in this important field of engineering. Because the books are designed not only to provide a grounding in the theory but to cover the range of applications, changes in protection technology mean that a process of updating is required. The second edition therefore presented a substantial revision of the original material and, although only minor changes have been made to the first three volumes, the publication of the fourth book in 1995, along with revised versions of existing works, reflects the considerable developments in the field of digital technology and protection systems.

The four revised volumes comprise 23 chapters, each with a bibliography. The aim remains that of providing sufficient knowledge of protection for those concerned with design, planning, construction and operation to understand the function of protection in those fields, and to meet the basic needs of an engineer intending to specialise in the subject.

In the use of symbols, abbreviations and diagram conventions, the aim has been to comply with British Standards.

The Electricity Training Association wishes to acknowledge the work both of the original authors, and of those new contributors who undertook the work of revising the first three volumes for this new edition; they are referred to at the head of the appropriate chapter. The Association also acknowledges the valuable assistance of National Grid Company, East Midlands Electricity,

Norweb, and Yorkshire Electricity staff and the help of the following in permitting reproduction of illustrations and other relevant material:

Allen West Company Limited, ASEA Brown Boverie Company Limited, BICC Limited, Electrical Apparatus Company, ERA Technology Limited, GEC Switchgear Limited, GEC Transformers Limited, Price and Belsham Limited, Reyrolle Protection Limited.

Extracts from certain British Standards are reproduced by permission of the British Standards Institution from whom copies of the complete standards may be obtained.

A particular indebtedness is acknowledged to the Chairman and members of the Editorial Panel, who directed and co-ordinated the work of revision for publication.

Chapter authors

R.K. Aggarwal	B Eng, PhD, C Eng, MIEE, SMIEEE
N. Ashton	C Eng, FIEE, ARTCS
K.A.J. Coates	C Eng, MIEE
P.C. Colbrook	BSc Tech, AMCST, C Eng, MIEE
L. Csuros	Dipl Eng, C Eng, FIEE
D. Day	BSc, C Eng, MIEE
P.M. Dolby	C Eng, MIEE
L.C.W. Frerk	BSc (Eng), C Eng, MIEE, FI Nuc E
F.W. Hamilton	B Eng, C Eng, FIEE
J. Harris	C Eng, MIEE
J.W. Hodgkiss	MSc Tech, C Eng, MIEE
L. Jackson	BSc, PhD, C Eng, FIEE
G.S.H. Jarrett	Wh Sc, BSc (Eng), C Eng, FIEE
M. Kaufmann	C Eng, FIEE
E.J. Mellor	TD
K.G. Mewes	Dip EE, C Eng, MIEE
P.J. Moore	B Eng, ACGI, PhD, C Eng, MIEE
J.H. Naylor	BSc (Eng), C Eng, FIEE, AMCT, DIC
C. Öhlén	MSc Electrical Engineering
H.S. Petch	BSc (Eng), MIEE, M Amer IEE
J. Rushton	PhD, FMCST, C Eng, FIEE
E.C. Smith	AMCT, C Eng, FIEE
C. Turner	DSc, F Inst P
H.W. Turner	BSc, F Inst P
J.C. Whittaker	BSc (Eng), C Eng, FIEE

Editorial panel

Protection symbols used in circuit diagrams

Fuse

Link – readily separable contact

Link – bolted contacts

Link – hinged or sliding

Link – plug-in type

Plug and socket

Auxiliary switch or relay contacts

Make contact

Break contact

Make contact with delayed make

Break contact with delayed break

Changeover contact

Push button switches

Make contact

Break contact

Control or selector switch

T N C

Note: the position of the rectangle represents the position in which the circuit is completed between the associated terminals

Circuit breaker

Circuit breaker normally open

Withdrawable metal-clad circuit breaker

Switch disconnector

Centre rotating post disconnector

Telephone type relay contacts

Make contact unit

Break contact unit

Changeover (break before make) contact unit

Changeover switch break before make

Make contactor

Mechanical coupling

Example – double pole contactor

Indirectly-heated bimetallic thermal element

Operating coil for contactors and relays – general

Coil with flag indicator

Series coil

Machine windings

General and shunt

Series

AC generator

Motor

Core (*if desired to indicate*)

Transformer or reactor winding

Single-break disconnector or earth switch

Additions to symbol for power-operated disconnector:
NA   – Non-Automatic
A   – Automatic

Fault throwing switch

Summation current transformer

Power and voltage transformers

Two winding

Simplified form

Auto-transformer

Simplified form – with delta tertiary

Transductor

Spark gap

Protective gas discharge tube

Capacitor

Power system protection

Current transformer

Current transformer with tapping

Interposing transformer current or voltage

Fixed resistor

Variable resistor

Resistor, with non-linear current/voltage characteristic

Impedance

Impedance, with non-linear current/voltage characteristic

Earthing resistors

Dry type

Liquid type

Arc suppression coil

Earth

Rectifier

Carrier-coupling equipment

Line

To carrier equipment

Fault

pn diode or semiconductor rectifying diode

Zener diode

Thyristor

pnp transistor

npn transistor

† *Envelopes may be omitted*

Thermionic valve, triode, indirectly heated

Cold cathode discharge tube (e.g., neon lamp)

Cold cathode trigger tube

Coaxial line

Cable sealing ends

Symbol	Description
	Rectifier equipment in bridge connection
	Amplifier
B	Buchholz – single float
B	Buchholz – two float
WT	Winding temperature single switch
WT	Winding temperature double switch
HSA	High speed ammeter
	Alarm flag relay
	Trip flag relay
R	Relay – general symbol
FP	Feeder protection
PLC PC	Power line carrier phase comparison protection
Z	Distance protection
MHO	High speed distance (Mho) protection
	Electric bell
	Signal lamp

Symbol	Description
E	Electromotive force (emf)
PP	Private pilot protection
POP	Post Office pilot protection
TP	Transformer protection
DB	Biased differential protection
PB D	Plain balance differential protection
T HVC	Transformer HV connection protection
CC	Circulating current
BB	Busbar protection
MCP	Mesh corner protection
HAR	High speed auto-close relay
OV	Overvoltage relay
X	High speed distance (reactance) protection (inverse definite minimum time)
3OC I	Three-pole overcurrent relay (inverse definite minimum time
2OC I EI	Two-pole overcurrent and single pole earth fault relayse (inverse definite minimum time)

OC I	One-pole overcurrent relay (inverse definite minimum time)
EI	Earth-fault relay (inverse definite minimum time)
3OC 2S I	Three-pole two stage overcurrent relay (inverse definite minimum time)
3OC DI	Three-pole directional overcurrent relay (inverse definite minimum time)
3OC	Three-pole overcurrent relay (instantaneous)
3OC XI	Three-pole overcurrent relay (extremely inverse definite minimum time)
E	Earth-fault relay (instantaneous)
3OC HS	Three-pole high set overcurrent relay
E SB LTI	Standby earth-fault relay (long time inverse definite minimum time)
E2S SB LTI	Two-stage standby earth-fault relay (long time inverse definite minimum time)
RP	Reverse power relay
E RES	Restricted earth-fault relay
T	Tripping relay
INT	Intertrip relay
INT S	Intertrip relay (send)
INT R	Intertrip relay (receive)
TD	Definite time relay
NEG PH SEQ	Negative phase sequence
LE	Lost excitation
3OC VC I	Three-pole voltage-controlled overcurrent relay (inverse definite minimum time)
3OC INT I	Three-pole overcurrent interlocked relay (inverse definite minimum time)
BF C CK	Breaker fail current check
LVC	LV connection protection

Chapter 8

Overcurrent protection

by J.W. Hodgkiss

8.1 Introduction

When the first small power systems were set up, the need to add automatic protection was soon realised. Equipment responsive to excess current (in the first place by fuses) was the obvious solution to the difficulties which had arisen and still today, by far the majority of all circuits are protected by this means. Selective action was soon needed and the graded overcurrent system has evolved to give discriminative fault protection.

Overcurrent protection should not be confused with overload protection which is related to the thermal capability of plant or circuits, whereas overcurrent protection is primarily provided for the correct clearance of faults. Very often, however, settings are adopted which make some compromise in order to cover both of these objectives.

Overcurrent protection is achieved by the use of fuses, by direct-acting trip mechanisms on circuit breakers or by relays.

8.2 Types of overcurrent system

Where a source of electrical energy feeds directly to a single load, little complication in the circuit protection is required beyond the provision of an overcurrent device which is suitable in operating characteristics for the load in question, i.e. appropriate current setting possibly with a time-lag to permit harmless short time overloads to be supplied.

Development of the power-system to one such as is shown in Fig. 8.2, in which the power source A feeds through a number of substations B, C, D and E, from each of the busbars of which load is taken, necessitates a more selective treatment. It is usually not good enough to shut down the whole system for every fault along the line.

The system requires discriminative protection designed to disconnect the minimum amount of circuit and load that will isolate the fault. Circuit breakers are installed at the feeding end of each line section and a graded scheme of protection is applied.

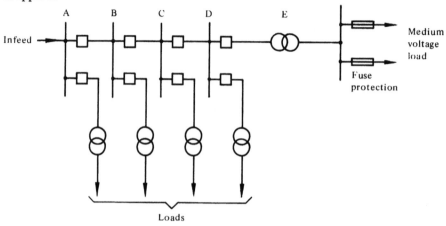

Fig. 8.2   *Radial distribution system*

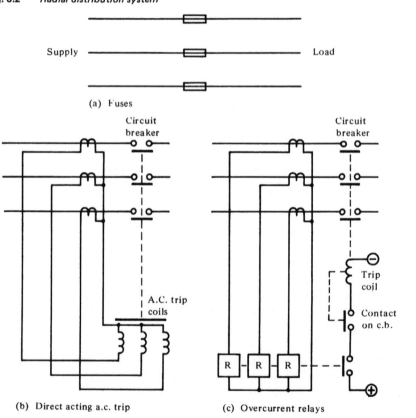

Fig. 8.2.1A   *Types of overcurrent protection*

8.2.1   Overcurrent and earth-fault protection systems

Overcurrent protection involves the inclusion of a suitable device in each phase, since the object is to detect faults which may affect only one or two phases. Where relays are used, they will usually be energised via current-transformers, which latter are included in the above statement. Typical arrangements are shown in Fig. 8.2.1A

The magnitude of an interphase fault current will normally be governed by the known impedances of the power plant and transmission lines; such currents are usually large.

Phase-to-earth fault current may be limited also by such features as:

(a)   the method of earthing the system neutral

(b)   the characteristics of certain types of plant, e.g. faults on delta-connected transformer windings

(c)   resistance in the earth-path.

In consequence, earth-fault current may be of low or moderate value and also often rather uncertain in magnitude particularly on account of item (c) above.

The protection is often required to have a high sensitivity to earth faults, i.e. earth-fault settings are often required to be lower than system rating. Response to an earth fault at a lower current value than system rating or loading is achieved by the residual connection shown in Fig. 8.2.1B. Three current-transformers, one in each phase, have their secondary windings connected in parallel and the group connected to a protective device, either circuit-breaker tripping coil or a relay.

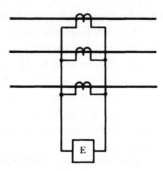

Fig. 8.2.1B   *Residual circuit and earth-fault relay*

With normal load current, the output of such a group is zero, and this is also the case with a system phase-to-phase short circuit. Only when current flows to earth is there a residual component which will then energise the protective device.

Since the protection is not energised by three-phase load current, the setting can be low, giving the desired sensitive response to earth-fault current. Phase-fault and earth-fault protection can be combined, as shown in Fig. 8.2.1C.

The protection scheme, involving equipments at a series of substations, can be graded in various ways, the significance of which is examined below.

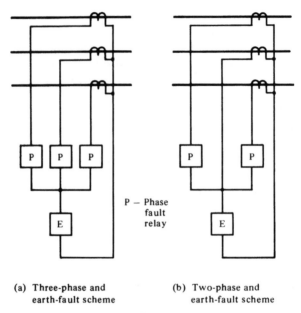

(a) Three-phase and
      earth-fault scheme

(b) Two-phase and
      earth-fault scheme

Fig. 8.2.1C   *Combined phase and earth-fault protection*

8.2.2 Grading of current settings

If protection is given to the system shown in Fig. 8.2 by simple instantaneous tripping devices, set so that those furthest from the power source operate with the
lowest current values and progressively higher settings apply to each stage back
towards the source, then if the current were to increase through the range of
settings, the device with the lowest setting of those affected would operate first
and disconnect the overload at the nearest point. Faults, however, rarely occur in
this way; a short circuit on the system will immediately establish a large current of
many times the trip settings likely to be adopted, and would cause all the tripping
devices to operate simultaneously.

The position would appear to be better when the feeder sections have sufficient
impedance to cause the prospective short-circuit current to vary substantially over
the length of the radial system, as indicated in Fig. 8.2.2. One might attempt to set
the circuit breaker trips to just operate with the expected fault current at the end
of the associated feeder section, but this would not be successful since:

(i)   It is not practicable to distinguish in magnitude between faults at F_1 and F_2,
       since these two points may, in the limit, be separated by no more than the
       path through the circuit breaker. The fault current at the alternative fault
       locations will then differ by only an insignificant amount (e.g. 0·1% or less)
       necessitating an unrealistic setting accuracy.

(ii)   In the diagram, the fault current at F_1 is given as 8800 A when the source
       busbar fault rating is 250MVA. In practice the source fault power may vary

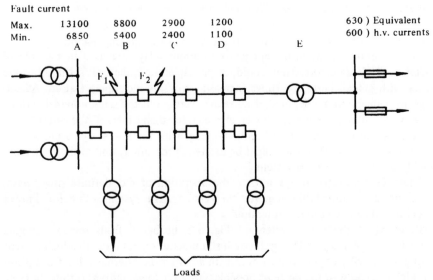

Fault current						
Max.	13100	8800	2900	1200		630) Equivalent
Min.	6850	5400	2400	1100		600) h.v. currents
	A	B	C	D	E	

Loads

Fig. 8.2.2   *Radial system with variation in fault current due to feeder impedance*

over a range of almost 2:1 by, for example, the switching out of one of two supply transformers; in some cases a bigger range is possible. A minimum source power of 130MVA may be assumed for illustration in this example, for which the corresponding fault current at F_1 would be only 5400 A and for a fault close up to Station A the current would be 6850 A. A trip set at 8800 A would therefore not protect any of the relevant cable under the reduced infeed conditions.

It is, therefore, clear that discrimination by current setting is not in general feasible.

A single exception to the above deduction occurs when there is a lumped impedance in the system. Where, as in the last section of Fig. 8.2.2, the line feeds a transformer directly, without other interconnections, there will be a significant difference in the fault current flowing in the feeder for faults, respectively, on the primary and on the secondary side of the transformer. In many cases, it is possible to choose a current setting which will inhibit operation for all secondary side faults, while ensuring operation for all primary side faults under all anticipated infeed conditions. The transformer is not necessarily adequately protected by such a high set protection; the protection of the transformer will be discussed later.

In special cases, instantaneous relays with high settings are used as a supplementary feature to other protection systems. This subject is described in more detail in Section 8.3.5.

8.2.3   Grading of time settings: the definite-time system

The problem discussed in the preceding Section is resolved by arranging for the equipment which trips the circuit breaker most remote from the power source, to operate in the shortest time, each successive circuit breaker back towards the supply station being tripped in progressively longer times; the time interval between any two adjacent circuit breakers is known as the 'grading margin'.

In this arrangement, instantaneous overcurrent relays are used in the role of starters or fault detectors. They could, in principle, have identical settings but are better with graded current settings increasing towards the supply station. All settings are subject to a tolerance; if all relays were given the same nominal setting, some would in fact operate at a lower current value than others. If the current were to be raised into the tolerance band, the relay with the lowest operating current would operate first. This relay might be located at the supply end of the feeder and would shut down the entire system.

These fault-detector relays initiate the operation of d.c. definite time relays, which are set to provide the required time grading. This system is therefore known as 'definite-time overcurrent protection'.

Referring to the radial system of Fig. 8.2, busbar E feeds separate circuits through fuses. A relay at D might have instantaneous operation with a high current setting which will not permit operation with a fault at E. Alternatively, if this is not feasible on account of the range of possible current, a lower current setting, but one above the maximum load current, may be used with a time setting chosen to discriminate with the fuse blowing.

Usually a time-lag of 0·2 s is sufficient, although it is desirable to check the suitability of this value for a fault current equivalent to the overcurrent element setting.

The relay at C may be set to operate in 0·5 s longer than that at D, i.e. in 0·7 s, and those at B and A will be progressively slower by the same amount, giving an operating time for relay A of 1·7 s.

Since the timing is not related to fault current, but is based only on position, the difficulty that was discussed for current grading does not exist here. A fault at any point will be removed by tripping the nearest circuit breaker on the supply side, which will occur before any of the others which carry the fault current have time to operate. The minimum amount of power system is thereby isolated, although for a fault on any but the last section, some disconnection of unfaulted sections and loss of load is inevitable.

The only disadvantage with this method of discrimination is that faults close to the power source, which will cause the largest fault current are cleared in the longest time.

At substations B, C and D, loads are connected through transformers. The time setting on these circuits are selected in the same way as that at D on the feed to E and should never be greater than that on the outgoing feeder from the same busbars.

8.2.4   Grading by both time and current: inverse-time overcurrent systems

The disadvantage of the definite time system mentioned above, is reduced by the use of protective devices with an inverse time-current characteristic in a graded system. Fig. 8.2.4 shows two time-current curves in which the operating time is in inverse ratio to the excess of current above setting. An exact inverse ratio has been used in this diagram in order that it may be seen that the effects obtained in the grading are general and not related to any specific device other than that the device has an inverse type of characteristic.

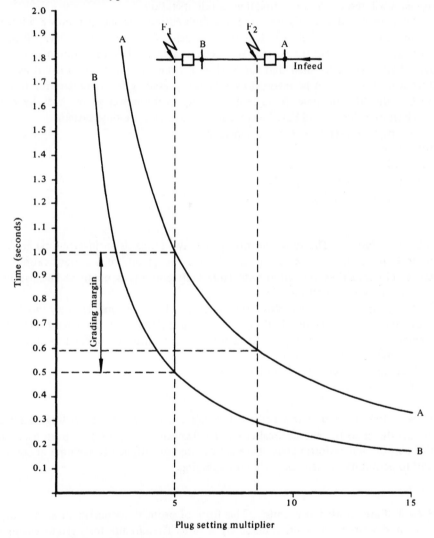

Fig. 8.2.4 *Principle of inverse time-grading*

The two curves correspond to protective devices having the same current setting, whilst the curve A shows twice the operating time as does curve B for any current value. The current scale is in multiples of setting current.

Two sections of a radial system A-B- etc. protected at A and B by the corresponding devices, is shown with a fault in alternative positions immediately after stations A and B.

Fault F_1 following station B produces 5 times setting current in both devices which tend to operate in 0·5 and 1·0 s, respectively. There is therefore a 'grading margin' of 0·5 s whereby the fault can be expected to be cleared at B and protective device A will reset without completing its full operation.

Alternatively, if fault F_2 occurred, the fault current through A is now 8·5 times setting, and device A will operate in 0·59 s which is little slower than device B with close-up fault F_1. Had the current step been greater, the best operation of A might have been even faster than that for B, although such a gain in speed is unusual in practice. This gain can be made at every stage of grading for a multisection feeder so that the tripping time, for a fault close up to the power source, may be very much shorter than would have been possible with a definite-time system.

Inverse-time overcurrent systems include:

(*a*)   fuses
(*b*)   delayed action a.c. trip coils
(*c*)   fuse shunted a.c. trip coils
(*d*)   inverse-time relays.

8.2.4.1   Fuses: The fuse, the first protective device, is an inverse-time graded protection. Although fuses are graded by the use of different current ratings in series, discriminative operation with large fault currents is obtained because the fuses then operate in different times.

This continues to be effective even when the clearance times are very short, because the fuse constitutes both the measuring system and the circuit breaker, whereas with a relay system, a fixed time margin must be provided to allow for circuit-breaker clearance time.

Fuses are dealt with fully in Chapter 5.

8.2.4.2 Delayed action trip coils: Grading can be achieved by fitting delaying devices direct to the circuit-breaker trip mechanism. Dash-pots have been used for this purpose but unfortunately these devices are not sufficiently accurate or consistent to permit more than relatively crude grading.

8.2.4.3 Fuse-shunted trip coils: This form of protection consists of an a.c. trip coil on the circuit breaker, shunted by a fuse. The circuits for overcurrent and combined overcurrent and earth fault are shown in Fig. 8.2.4.3A.

The resistance of the fuse and its connections is low compared with the impedance of the trip coil, and consequently the majority of the current, supplied by the current transformer, passes through the fuse. If the current is high enough when a fault occurs, the fuse blows and the current is then transferred to the trip coil. The time taken, from the instant when the fault occurs, for the circuit breaker to trip is thus mainly dependent on the time/current characteristics of the fuse.

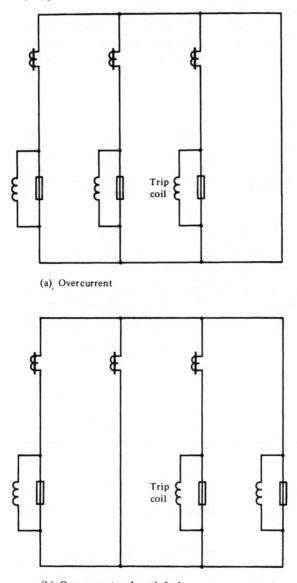

(a) Overcurrent

(b) Overcurrent and earth-fault

Fig. 8.2.4.3A   *Arrangements for fuse-shunted trip coils*

Some error is introduced by the current which is shunted from the fuse through the trip coil, but this can be made small. The coil and its associated circuit should have an impedance of many times that of the smallest fuse that is to be used. The impedance of the coil must be low enough, however, to not unduly overburden the current transformer, when the fuse blows.

It is now standard practice to employ 2 A trip coils with current transformers of 5 A secondary rating and to use 2·5, 5·0, 7·5 or 10 A fuses as required. The rating of a time-limit fuse is generally specified in terms of the minimum current required to cause the fuse to blow, whereas in the usual application of a fuse, its rating is the maximum current which can be carried continuously without deterioration.

In the design of the circuit, the fuses should be connected directly, not through isolating plugs, to the current transformers, to avoid the inclusion of additional resistance in the fuse circuit.

Fig. 8.2.4.3B shows typical time/current characteristics for time-limit fuses. It will be noted that the difference in operating time between any two ratings of fuse

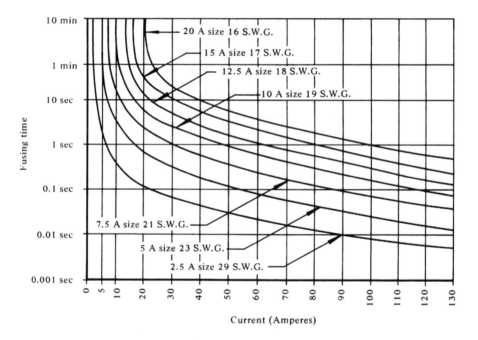

Fig. 8.2.4.3B   *Time/fusing current characteristics of time-limit fuses*

decreases as the current increases. These differences constitute the discrimination time. Unlike main fuse protection, this margin must not be reduced to a very low value, since it must cover the tripping time of the circuit breaker. Thus for 0·5 s margin the fault current must not exceed a certain limit. For example, for a current of 37 A there is 0·5 s between the time for blowing a 5 A and a 7·5 A fuse. If these fuses were used in conjunction with 300/5 A current transformers, for the protection of two sections of a radial feeder, the maximum fault current at the beginning of the second section, that is at the sending end of the second feeder, must not exceed

$$\frac{37 \times 300}{5} = 2220A$$

If the maximum fault current at this location could be higher, it would be necessary to use a larger rating for the first section fuse. It would then be necessary to decide whether this is admissible under considerations of minimum fault current, or of desired overload protection.

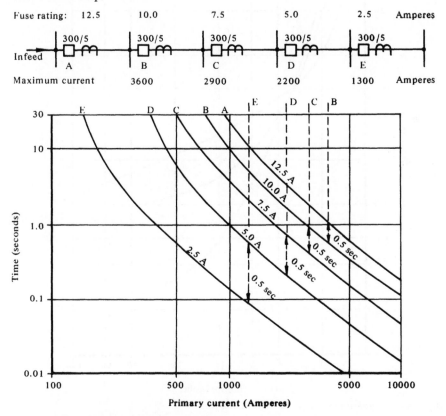

Fig. 8.2.4.3C   *Application of time-limit fuses*

As an example in the application of this form of protection, Fig. 8.2.4.3C shows a radial feeder, using graded time-limit fuses, and shows in graphical form the characteristics of the fuses and the discrimination obtained. The current scale of primary current allows for the current-transformer ratios, in this case uniformly of 300/5 A.

In practice, the limits imposed by the maximum permissible fault currents preclude a more extensive use of this form of protection. In addition, the fuses have not proved to be entirely satisfactory; their characteristics can be changed by through-fault current and also by ageing. In view of this, all fuses that have carried fault current even if apparently sound, should be replaced with new fuselinks.

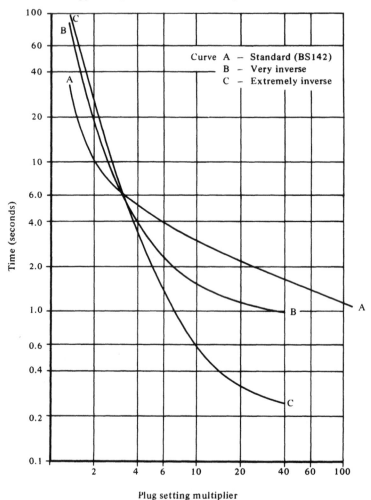

Plug setting multiplier

Fig. 8.2.4.4   *Typical time curves for IDMTL relays*

8.2.4.4 Inverse-time overcurrent relays: Relays designed to provide an inverse characteristic are both more reliable and provide more scope for precise grading than the above techniques. The relays are usually, although not inevitably, of the induction disc type and are adjustable over a wide range in setting current and operating time. A description of relay design details for this application is given in Volume I, Chapter 6, Section 6.2.3.

Typical current-time characteristics are shown in Fig. 8.2.4.4. Curve A is the most commonly used characteristic and is standardised by BS 142: Parts 1-4, 1990 (1993).

It will be noted that the curve, although nominally inversive, departs from an exact ratio at each end due to the effects of mechanical restraint and saturation of the electro-magnet. Steeper characteristics curves B and C are known as 'very inverse' and 'extremely inverse' characteristics, respectively, and have special applications.

It will be observed that the characteristic curves are plotted with an abscissa of 'plug setting multiplier'. Relays can have many different settings for which there are as many time-current curves. This cumbersome state of affairs is simplified by realising that a relay of a given type always has the same ampere-turn loading of its winding at setting and for each operating time point, the coil turns being in inverse ratio to the setting. It follows that a curve plotted in terms of multiples of setting current is applicable to all relays of that type and at all plug settings. Taking into account the current transformer with which the relay is used we can write:

$$\text{plug setting multiplier (p.s.m.)} = \frac{\text{primary current}}{\text{primary setting current}}$$

The ordinate is the operating time scale for maximum disc movement. Restricting the movement by setting forward the back stop proportionately reduces the operating time, the ratio of such reduction applying at all current levels and hence being known as the time multiplier setting (t.m.s.). When the curve is plotted with logarithmic scales, the effect is to displace the basic curve downwards, maintaining the original curve shape. Because of certain small errors in the constancy of the time multiplier over the current range, a family of curves with t.m.s. from 0·1 to 1·0 is sometimes given by the relay manufacturer.

The principle of grading is generally similar to that of the definite-time system. For any fault, the relays in a series system are graded so that the one nearest to the fault trips the associated breaker before the others have time to operate. A grading margin of 0·4 or 0·5 s is usually sufficient.

8.3   Selection of settings

It will be clear from the generalised discussion above that a knowledge of the fault current that can flow is essential for correct relay application. Since large-scale power-system tests are normally impracticable, fault currents must be calculated. It is first necessary to collect system data and then to calculate maximum and minimum fault currents for each stage of grading. Time-grading calculations are made using the maximum value of fault current; the grading margin will be increased with lower currents so that discrimination, if correct at the highest current, is ensured for all lower values.

Minimum fault currents (i.e. short-circuit current with minimum amount of feeding plant or circuit connections) are determined in order to check that current settings are satisfactory to ensure that correct operation will occur.

8.3.1   System analysis

Some or all of the following data may be needed:
(*a*)   A one-line diagram of the power system, showing the type of all protective devices and the ratio of all protective current transformers.
(*b*)   The impedances in ohms, percent or per unit, of all power transformers, rotating machines and feeder circuits.

It is generally sufficient to use machine transient reactance X'_d, and to use the symmetrical current value; subtransient effects and offset are usually of too short duration to affect time-graded protection.
(*c*)   The starting current requirements of large motors and the starting and stalling times of induction motors may be of importance.
(*d*)   The maximum peak load current which is expected to flow through protective devices. 'Peak load' in this context includes all short-time overloads due to motor starting or other causes; it does not refer to the peak of the current waveform.
(*d*)   Decrement curves showing the rate of decay of the fault current supplied by generators.
(*f*)   Excitation curves of the current transformers and details of secondary winding resistance, lead burden and other connected burden.
Not all the above data is necessary in every case; with discretion some items may be deemed to be irrelevant.

The maximum and minimum values of short-circuit currents that are expected to flow through each protective device are calculated. Three-phase short-circuit calculations are adequate for phase-fault studies and are relatively simple.

Earth-fault current and its distribution through the system should also be examined. This will be necessary if the system is earthed through a limiting impedance, in cases of multiple solid earthing, and also if the ratio of phase-fault current to overcurrent setting is not large. Earth-fault calculations will be required whenever earth-fault relays are used, but should be made also in the above cases, even if only overcurrent elements are included, since in any case earth faults will have to be covered by the protection.

The calculation of fault current is covered in Volume 1, Chapter 3.

8.3.2   Grading of relay settings

Certain guiding principles apply to the design of a graded system of protection;
(*a*)   Wherever possible, in any graded sequence, use relays having the same operating characteristic at all points.

This is not an absolute rule but if this practice is not followed, extra care is necessary to ensure that discrimination is maintained at all current values.
(*b*)   Choose current settings for all relays. This choice is arbitrary to a large extent, but must take into account maximum load currents and legitimate

short-time overloads caused by, for example, the starting of large motors.

Setting currents as above are related to primary system currents. Most relays, however, are connected to the system through current transformers, and the combination is to be regarded as a single entity. Thus a relay having a rated current of 5 A and a setting of 150% would have a current setting of 7·5 A. If the relay were operated from a current-transformer having a ratio of 300/5 A, it would have a primary setting of

$$7·5 \times \frac{300}{5} = 450 \text{ A}$$

Provided the rated current of the relay is identical with the secondary rated current of the current transformer, the percentage setting can be applied directly to the primary rating of the c.t.; hence, also in the above example,

$$\text{primary setting} = \frac{150}{100} \times 300 = 450 \text{ A}$$

Where the choice of relay setting is referred to below, it is the primary setting which is implied.

Primary setting currents should be graded so that the relay furthest from the power source has the lowest setting and each preceding relay back towards the source has a higher setting than that following. Not only does this ensure that relay and c.t. errors do not produce a range of current value within which maldiscrimination may occur, but also it allows for load which may be taken from the intermediate substations.

(c) Time grading for inverse time-current systems should be calculated at the highest possible fault current for each grading stage. The grading margin will be greater at lower currents.

(d) The 'grading margin' is required to cover:
   (i)   Circuit-breaker clearance time
   (ii) Relay overshoot
   (iii) Relay and current-transformer errors
   (iv) An allowance to ensure a final contact gap for the discriminating relay
   See Volume 1, Chapter 6, Section 6.2.3 for further discussion.

A margin of 0·4 s is sufficient with modern switchgear and relays. Old, slower and less accurate equipment may need a little more.

8.3.2.1 Grading for definite-time relays: The overcurrent starting elements must be given settings, as stated above, which are higher than the greatest peak loads. Moreover, if the element has a returning ratio (ratio of resetting to operating current) which is appreciably below unity, the resetting value should exceed the peak load. This is because the relay is liable to be operated by transformer inrush current or quickly cleared through-fault current; if following the current surge the relay holds in with sustained load current, the associated timing relay will continue to be energised and discrimination will be lost.

Settings should not be higher, however, than is necessary to comply with the above requirement with a reasonable margin. The assessment of maximum load of most distribution systems is based on a large diversity factor. Abnormal circumstances occasionally occur which cause a much higher proportion of available apparatus to be connected than is usual, leading to an excessive overload. The distribution system can carry a moderate overload for a considerable time by virtue of the thermal time-constant of lines, cables and transformers, but it is desirable for heavy sustained overloads to be tripped. The overload protection which can be obtained from the overcurrent relays, if these are carefully set, is a useful additional feature. The settings must also be below the minimum value of fault-current for a short circuit at the remote end of the relevant feeder section, and preferably below that for a short circuit at the end of the next following section. This latter condition enables each equipment to provide the very useful feature of 'back-up' to the next section in the event of any failure in that section to clear a fault.

There will usually be no difficulty in complying with this latter requirement, and it is well to leave as much margin below the calculated minimum current as is permitted by the peak load requirement. Although most faults are short circuits, this is not always so; relatively high-resistance faults do occur and although this is a subject of much uncertainty, a liberal operating margin will give assurance of good operation.

Time grading for this system consists simply in choosing an operating time for the last relay in the sequence to discriminate with lower voltage equipment such as distribution fuses (e.g. 0·2 s) and then adding a grading margin of 0·4 or 0·5 s at each stage back to the source station. With five relays in sequence, the supply station relay will then be delayed 1·8 - 2·2 s.

8.3.2.2 Grading for inverse-time relays: The principles set out above for definite-time protection are generally applicable here also but the following special points should be noted.

(*a*)   The improvement in time grading discussed in Section 8.2.4 is most marked when the relay is operating under conditions corresponding to the steeper portions of the characteristic curve, that is with a low value of plug setting multiplier. There is, therefore, a conflict of requirements, in that to obtain a low p.s.m. may involve the use of a high setting, which removes the possibility of obtaining even rough overload protection. A decision may be necessary as to whether overload protection or the best fault protection is required. In a particular case it may be necessary to provide relays having long inverse time lags and low settings for overload protection, and separate overcurrent relays graded to give the shortest fault-clearance times.

(*b*)   The standard inverse-time overcurrent relay does not operate with a current equal to its setting value but requires slightly more. In principle, an operating characteristic in which the time is inverse to the excess of current above setting current should asymptote to the setting current value, the operating time for setting current being infinite. The relay, however, cannot operate infinitely slowly since friction, ignored in the basic theory, would become dominant, in this condition. It is therefore accepted that the relay will not operate with setting current.

British Standard 142, Parts 1-4, for relays specifies that 'the relay shall not operate at a current equal to or less than the setting; the minimum operating current shall not exceed 130% of the setting.* Typical designs of modern relays will operate with about 115% of the setting in a fairly long time.

Relays with the steeper characteristics ('very inverse' and 'extremely inverse') will in general require a larger margin, for example 130% of setting current, to operate. This is because these relays usually have a lower operating torque, so that pivot friction is of more significance.

(c)   The restrictions on minimum current setting are alleviated, as compared with definite-time relays, by the fact that induction relays have a fairly high returning ratio. In fact, if following partial operation on through fault, the load remained high and approaching relay setting so as to inhibit resetting of the relay disc, discrimination would still not be lost since the induction relay has to be driven forward by current above its setting value to complete operation.

Grading can now be considered for the radial system shown in Fig. 8.3.2.2, in which power is supplied from a source point A to substations B, C, D and E at each of which loads are fed through step-down transformers.

The range of fault current must first be established. Maximum fault current for a three-phase short circuit at each station in turn, with the system fully connected, is calculated. In this example it is convenient to express the 33kV system and the 33/11kV transformers as percent impedance on a base of 100 MVA to obtain the fault power at A. The equivalent star impedance behind A is determined, to which value the impedance of the following feeder sections is added and the fault current calculated at each subsequent station.

Minimum fault current is calculated for the condition of one 10 MVA supply transformer being disconnected. A reduction of fault MVA in the 33kV feed might also have been considered but this has not been done in this case since:

(a)   A simultaneous outage in the 33kV and llkV systems is a double contingency with the fault in the 11kV system making a triple contingency. This should be covered, although some loss of performance may be tolerated, depending upon the probability of this event.

(b)   The effect is relatively unimportant in this example. Switching out one of the two parallel 33kV lines would not reduce the fault MVA on the 33kV busbar to half its normal value, i.e. would not double the source impedance, since some impedance must exist beyond these lines. The minimum fault current at busbar A would be reduced by perhaps 5% with less reduction for more remote faults. This produces only unimportant changes in the performance.

In other instances such a condition might be more significant and should then be assessed. The prospective fault currents are shown in Table 1.

*Quoted from BS142:1953; the 1966 edition contains the same data in a different presentation.

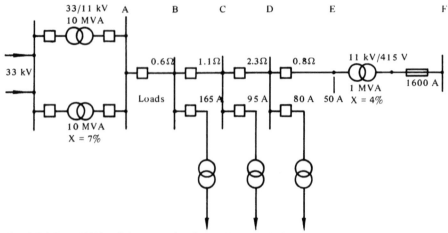

Fig. 8.3.2.2   *11kV radial system; data for grading calculation*

Maximum load current taken from each station is shown on Fig. 8.3.2.2; making the safe assumption that such peak loads are coincident and of similar power factor, the values are summated to give the feeder section loads. Current-transformer ratios have been previously chosen. Relay settings are now chosen so that primary settings have a sufficient margin over normal peak loads to permit some abnormal variation of load, whilst still ensuring that a degree of overload protection is provided. These values are all included in Table 1.

TIME GRADING

Grading is begun at the most remote station by choosing a suitable relay operating time. No data is given of the arrangements beyond the 1 MVA transformer except that fuse protection is used. Assuming that a single low-voltage main fuse is used having a rating equivalent to about 60 A in the 11kV line, (e.g. if the transformer has a ratio of 11000/415 V, a fuse rated at 1600 A) a short circuit immediately following will melt this fuse in approximately 0·05 s. Allowing an equal time for arcing, the fault will be cleared in 0·1 s. Since only relay overshoot has to be added, a relay operating time of 0·15 s is sufficient for discrimination. Note that a small margin is permissible, since with only a single fuse, little loss occurs if the breaker actually fails to discriminate. If, however, the load is divided into two circuits, each separately fused at half the above rating, a fault on either would be cleared in 0·03 s, so that more than sufficient margin is provided by the above relay time to ensure full discrimination.

The maximum 11kV fault current for the fault beyond the transformer is 626 A, corresponding to a p.s.m. equal to

$$\frac{626}{125} s = 5.008 \text{ (say } 5.01)$$

Table 1

Location	Total impedance from source				Fault current A		Max. section load A	CT ratio	Relay setting	
	Percent on 100 MVA base		Ω							
	Min.	Max.	Min.	Max.	Max.	Min.			%	Primary Amps
33kV bus	6.67	6.67								
11kV bus										
C.B. A	41.67	76.67	0.5042	0.9277	12 596	6 846	390	400/5	125	500
C.B. B			1.104	1.528	5 753	4 156	225	300/5	100	300
C.B. C			2.204	2.628	2 882	2 417	130	200/5	75	150
C.B. D			4.504	4.928	1.410	1 289	50	100/5	125	125
Medium voltage Bus F (1)			10.144	10.568	626	601	50			

Notes:

(1) Values of impedance, fault current and load current at location F refer to fault or load conditions on medium voltage system as seen from 11kV system.

(2) Fault currents etc. are calculated to a higher accuracy than is necessary or strictly justifiable in order to avoid apparent anomalies due to approximations.

Referring to Fig. 8.2.44 gives a time for full disc travel of 4·3s. The smallest time multiplier setting which it is wise to use in normal circumstances is 0·05, which gives an actual time of

$$4·3 \times 0·05 = 0·215s$$

This value is greater than the value chosen above (0·15 s) and is therefore satisfactory to discriminate with the fuse.

A fault on section DE close to D will result in a maximum current of 1410A, corresponding to a p.s.m. of 11·3 and a full travel time (T_C) of 2·82s. The t.m.s. is 0·05 as before, so that the actual time (T_A) is $T_C \times$ t.m.s. = 2·82 x 0·05 =0·141 s. The next relay nearer to the source at C will carry the fault current and must discriminate. Allowing 0·5 s grading margin, this relay should have a prospective operating time of 0·641 s. Relay C has a p.s.m. $= \dfrac{1410}{150} = 9·4$ under this condition, for which the curve time is 3·1 s. Relay C must therefore be given a time multiplier setting $= \dfrac{0·641}{3.1} = 0·207.$

When a fault is placed close to Station C the fault current becomes 2882 A equivalent to a p.s.m. of 19·2 and a curve time of 2·23 s. The t.m.s. has already been chosen above to be 0·207 in order to grade with relay D, so that the operating time is now

$$T_A = 2·23 \times 0·207 = 0·462 \text{ s.}$$

The process of grading relays B and A is similar to the steps set out above; the complete grading calculation is set out in Table 2. It is well worthwhile, in any grading calculation, to arrange the working in a table of this form so that the results can be easily reviewed.

This example shows that the operating time for relay A is not the sum of the grading margins. Although three steps of grading, each with 0·5 s margin are provided, relay A operates for a close-up fault in 0·86 s; if a definite time system with a similar margin had been used the operating time for a fault close to the supply station A would have been 1·65 s, (i.e. 0·15 + 3 x 0·5 = 1·65 s).

The gain is due to the reduction in operating time for each relay as the fault is moved from the following station where its t.m.s. for grading is calculated, to the close-up position.

The calculation is performed for the maximum fault current relevant to each grading step; e.g. relay C is graded with D with a fault close to D, not further down the system at E or F. If grading is performed at the highest possible current, even greater margins will exist under lower current conditions. This fact is illustrated by

Table 2: *Maximum fault current*
(Relays with standard characteristic specified in BS 142)

Relay Location	Setting Primary (A) p.s.m.	Fault close to station:																		
		F				**D**				**C**				**B**				**A**			
		p.s.m.	T_C	t.m.s.	T_A	p.s.m.	T_C	t.m.s.	T_A	p.s.m.	T_C	t.m.s.	T_A	p.s.m.	T_C	t.m.s.	T_A	p.s.m.	T_C	t.m.s.	T_A
D	125	5.01	4.3	0.05	0.215	11.3	2.82	0.05	0.141												
C	150					9.4	3.1	0.207	0.641	19.2	2.23	0.207	0.462								
B	300									9.61	3.05	0.315	0.962	19.2	2.23	0.315	0.702				
A	500													11.5	2.8	0.43	1.202	25.2	2.0	0.43	0.86

Note: Some values in the above table have more significant figures than would be justifiable in practice; this is done so that the working and results are not confused by approximations.

p.s.m.   = plug setting multiplier
T_C = time for full travel from characteristic curve, s
t.m.s.   = time multiplier setting
T_A = actual time, i.e. $T_C \times$ t.m.s., s

Table 3: *Minimum fault current (one supply transformer disconnected) Relays with standard characteristic specified in BS 142*

Relay Location	Setting Primary A	F p.s.m.	F T_C	F t.m.s.	F T_A	D p.s.m.	D T_C	D t.m.s.	D T_A	C p.s.m.	C T_C	C t.m.s.	C T_A	B p.s.m.	B T_C	B t.m.s.	B T_A	A p.s.m.	A T_C	A t.m.s.	A T_A
D	125	4.81	4.4	0.05	0.22	10.31	2.96	0.05	0.148												
C	150					8.59	3.22	0.207	0.666	16.1	2.4	0.207	0.497								
B	300									8.06	3.3	0.315	1.04	13.9	2.6	0.315	0.82				
A	500													8.31	3.27	0.43	1.41	13.7	2.6	0.43	1.12
Time margin									0.518				0.543				0.59				

Note: Actual time margins are increased above the grading margin (0.5 s) used in Table 2

p.s.m. = plug setting multiplier
T_C = time for full travel from characteristic curve, s
t.m.s. = time multiplier setting (these values are taken from Table 2)
T_A = actual time, s

Table 4: *Maximum fault current*
Relays with 'very inverse' characteristic

Relay Setting		F				D				Fault close to station: C				B				A			
Location	Primary (A)	p.s.m.	T_C	t.m.s.	T_A	p.s.m.	T_C	t.m.s.	T_A	p.s.m.	T_C	t.m.s.	T_A	p.s.m.	T_C	t.m.s.	T_A	p.s.m.	T_C	t.m.s.	T_A
D	125	5.0	2.0	0.1	0.2	11.3	0.52	0.1	0.052												
C	150					9.4	0.66	0.84	0.552	19.2	0.34	0.84	0.29								
B	375									7.68	0.9	0.88	0.79	15.3	0.395	0.88	0.35				
A	800													7.9	1.0	0.85	0.85	15.7	0.39	0.85	0.33

Note:

p.s.m.	= plug setting multipler
T_C	= time for full travel from characteristic curve, s
t.m.s.	= time multiplier setting
T_A	= actual time, s

calculating the performance of each relay under the minimum fault currents quoted in Table 1. In this case, the t.m.s. values are those determined in Table 2. The results are shown in Table 3. It will be observed that reasonable operating times are still obtained and that each grading margin is slightly increased above 0·5 s.

8.3.2.3 Grading with 'very inverse' relays: Reference is made in Section 8.2.4.4 to alternative relay characteristics. The above example of grading is performed using the standard (BS142) characteristic, which is the one most commonly used. By steepening the characteristic, the recovery in total time, which is achieved by the speeding up of each relay as the fault position is moved through the relevant section of the system, is increased. In a typical case there may be little increase in total tripping time for the successive relays, back to the power source, whilst in an extreme case the tripping time may actually decrease in successive sections counting back to the supply station.

A grading calculation for the system of Fig. 8.3.2.2 and the maximum fault currents listed in Table 1 is set out in Table 4. Relays C and D have the same current settings as previously but the time multiplier setting of relay D has been increased to 0·1 in order to discriminate with the medium voltage fuse. The grading calculation then proceeds by similar steps to those in Table 2; it is found to be necessary to increase the current settings of relays B and A to 375 and 800 A, respectively, in order to provide a grading margin of 0·5 s.

It will be seen that the tripping time for a fault close to the supply station A is very substantially reduced, compared with that for the 'standard' relay. This reduction in time could be a very important gain in the operation of the power system.

It may well be asked why the very inverse relay is not used universally. One reason is that the steep characteristic is obtained at the expense of a considerable reduction in operating torque. The relay element is therefore more delicate in all respects, having a lower main contact rating, resetting more slowly, and obviously being more susceptible to failure when in a less than perfect state of mechanical condition and cleanliness.

The above comparison refers to electromechanical relays. If practice changes to the universal use of electronic relays, the position will require reassessment, taking into account all the features of the equivalent electronic relays.

8.3.2.4 Graphical method of grading: Fig. 8.3.2.4A illustrates the result of applying the settings calculated. Provided logarithmic scales are used, the effect of increasing the current setting is to shift the curve horizontally, and altering the time setting shifts the curve vertically. In neither case is there a change in shape. It is possible therefore to determine the required settings by using a transparent template cut to the shape of the relay time/current characteristic. Such a template is illustrated in Fig. 8.3.2.4B. The curve is plotted to the same scale as will be used

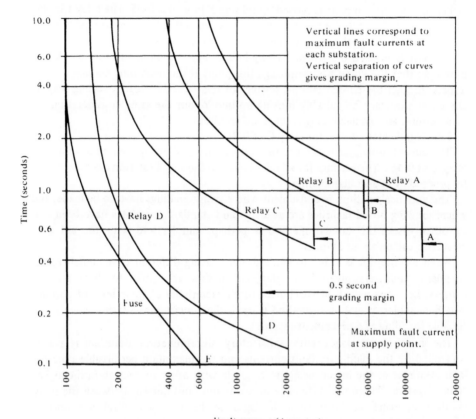

Fig. 8.3.2.4A   *Display of grading calculation for radial system (Table 2)*

for depicting the actual grading. In addition, a horizontal line X is drawn through the value 1·0 on the time scale to serve as an indicator line for the time multiplier, and the vertical line Y is drawn through the point p.s.m. = 1 to indicate the primary current setting. The grading operation is performed on a graph sheet with a similar sized ruling to that used for plotting the characteristic curve, but the abscisca is scaled in primary current over a sufficient range to accommodate the system fault-current values.

The maximum values of fault current at each relay position are calculated, and vertical lines are drawn through the corresponding scale values. Starting with the relay which is required to operate in the shortest time and at the lowest current (relay D in this example), place the template over the scales and adjust its position so that the vertical line Y corresponds to the current setting chosen with regard to load, and slide the template until the time at the current value for a fault at F is suitable to discriminate with the fuse protection at that point; that is, with 626 A, the time must be not less than 0·15 s. The horizontal line X then indicates the t.m.s. on the time scale. In this example, the t.m.s. indicated is 0·035, which is rather low, so the template is moved up till line X indicates 0·05. The relay tripping time is then 0·21. With the template in this position a line is drawn to represent the characteristic as far as the vertical line D, where the time is 0·14.

A point is marked 0·5 s higher along line D, i.e. at 0·64 s, and the template is moved so that the curve passes through this point at the same time making the line Y pass through the chosen current setting for relay C. The current setting may be chosen at this stage by suitably moving up line Y, but the same considerations of load current, and available plug settings are required.

Line X indicates a t.m.s. for relay C of 0·21.

The curve is now drawn as far as the vertical line C, showing a minimum time for relay C of 0·46s. The settings for all relays back to the supply station are determined by proceeding in a similar manner.

The graphical method is convenient where many gradings have to be made, and where drawing board facilities are available and standard templates have been cut. In other cases, the tabular method is perhaps simpler and with the aid of a calculator can be about as quick.

When an attempt is made to grade devices having different characteristics, e.g. 'standard' and 'very inverse' the graphical method may help to show possible dangers, i.e. unexpected encroachment of characteristics at some level of current. Whether tabular or graphical method is used, the concise presentation is clear and suitable for filing as a permanent record.

The number of stages with satisfactory discrimination between relays is controlled by the maximum fault currents and the maximum permissible time for fault clearance at the infeed position, to grade with other protection in the higher voltage system. Where the difference in maximum fault current between the relay points is appreciable, the number of stages can be increased within a permitted infeed clearance time, but where these differences are small the relays tend to operate as definite-time devices, whatever their basic characteristics, and the number of stages is more limited. When a limiting earthing resistor is used, there is likely to be comparatively little variation of earth-fault current throughout a radial system such as the above example. Fortunately, zero-sequence components do not pass beyond the usual delta-star transformer so that earth-fault current in a distribution system will produce only phase-phase currents in the higher voltage

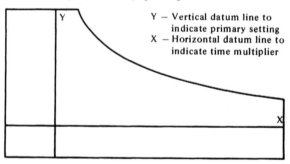

Fig. 8.3.2.4B   *Template for graphical grading method for IDMT relays*

system, which being of limited magnitude are less likely to involve the relevant protection. The time limitation which is imposed is likely to be less severe. Nevertheless, the number of earth-fault relay grading stages needs to be carefully examined.

8.3.3 Current transformer requirements

It is obvious that an inverse time/current relay cannot give its correct performance if it is not energised with the current for which its operating time is calculated. A relay and energising current transformer operate as a single entity, but the calibration of the relay is lost if the transformer does not faithfully reproduce the input current on the reduced scale. It is not easy to calculate an accurate correction for c.t. errors. An induction relay energised from a c.t. of inadequate output operates too slowly, partly because of ratio error in the c.t. and partly because of waveform distortion of the secondary current.

Where time/current grading is required the accuracy of the current transformer must be maintained up to many times the primary rating. This is possible with protective current transformers provided the total burden in VA at rated current of the secondary circuit, including relays, instruments if any and wiring is sufficiently below the output capability of the transformer. BS3938 classifies protective current transformers as 5P or 10P, corresponding to maximum errors of 5% and 10%, respectively, at the maximum current for measurement. The maximum current is specified in terms of an 'accuracy limit current' and the ratio of the accuracy limit current to the rated current is the 'accuracy limit factor'. Protective current transformers are specified in terms of VA at rated current, class and accuracy limit factor, e.g. 10 VA/5P/15.

The burden VA and the accuracy limit factor are approximately mutually inverse. If the secondary rating in the above is 5 A, then the voltage across the rated burden at rated current will be 2V, and will be 30V at the accuracy limit current. The latter value must be within the saturation limit of the c.t. secondary winding. If the burden is halved, the accuracy limit current can increase nearly but not quite to double the former figure, the lack of strict inverse proportionality being on account of the resistance of the secondary winding, which must be allowed for if an accurate assessment is required.

The following example illustrates the procedure:

impedance of burden (10VA at rated current)

$$= \frac{10}{5^2} \ = \ 0.4 \ \Omega$$

Suppose resistance of c.t. secondary winding is 0·15 Ω,

Total secondary impedance = 0·55 Ω.
(Arithmetical addition is usually good enough)
Secondary e.m.f. at accuracy limit current is 0·55 x 5 x 15 = 41·25 V,
which the c.t. must be able to develop.

If the burden is halved (i.e. to 0·2 Ω),

Secondary impedance = 0·2 + 0·15 = 0·35 Ω

$$\text{New accuracy limit current} = \frac{41 \cdot 25}{0 \cdot 35} = 117 \cdot 8 \text{ A}$$

New accuracy limit factor = 23·5

For higher burdens the same procedure gives a reduced factor. Provided that the total external burden is not greater than the rated value for the c.t. and the fault current up to which discrimination is required, expressed as a multiple of rated current, is not greater than the rated accuracy limit factor, then good performance can be expected.

If the above conditions are not satisfied, good results will still be obtained provided that the actual accuracy limit current computed as above for the actual total burden is not less than the relevant fault current. In the latter case there is no need to calculate a revised accuracy limit factor. In the case of the above c.t. rated 10VA/5P/15 with a secondary winding resistance of 0·15 Ω, it was seen that this specification required the c.t. to be able to develop a secondary e.m.f. of 41·25V without excessive saturation.

Supposing that this rating had been given for the c.t. at B in Table 1, it is seen that the required accuracy limit factor is

$$\frac{5735}{300} = 19 \cdot 18, \text{which is in excess of the rating.}$$

If on examination of the total external burden this can be reduced to 6 VA, then burden impedance

$$= \frac{6}{5^2} = 0 \cdot 24 \text{ Ω}$$

total impedance
= burden + c.t. winding

= 0·24 + 0·15 = 0·39

maximum allowable secondary current

$$= \frac{41 \cdot 25}{0 \cdot 39} = 105 \cdot 8 \text{A}$$

maximum allowable primary current

$$= 105 \cdot 8 \times \frac{300}{5} = 6346 \text{ A}$$

The allowable primary current exceeds the maximum fault current (5753A) and therefore the application will be satisfactory.

8.3.3.1   Burdens: A relay burden is quoted in VA at the setting. Relays of the same design but different settings will usually consume the same VA at their respective settings. This is also true when the various settings are obtained on the same relay by tapping the coil, since the same energy must be provided at setting, however many coil turns are used. Now one can write

$$\text{VA (burden)} = I^2 Z$$

where I is the setting current and Z is the coil impedance. Therefore

$$Z = \frac{VA}{I^2}$$

Thus with a 50% setting on a relay rated at 5 A, and with a declared burden of 3 VA:

$$Z_{50\%} = \frac{3}{2 \cdot 5^2} = 0 \cdot 48 \ \Omega$$

The burden at rated current is

$$5^2 \times 0 \cdot 48 = 12 \text{ VA}$$

Similarly, if the relay plug setting is 200%:

$$Z_{200\%} = \frac{3}{10^2} = 0 \cdot 03 \ \Omega$$

and the burden at rated current is:

$$5^2 \times 0 \cdot 03 = 0 \cdot 75 \text{ VA}$$

8.3.3.2   Variation of burden impedance: The above calculations are based on a declared burden at setting and are too simple if an accurate figure is required. The relay impedance deviates from these values in two ways, namely:

(a)   variation of VA burden with tap setting
(b)   variation of impedance with current value.

(a)   When a relay is used with a tap setting which is higher than the minimum of its range, only a portion of the coil is in use which is less efficient than is the whole coil. Thus the reactance of the coil will be reduced in the ratio of the square of the per unit turns, * but the resistance will only decrease approximately linearly with the coils turns. The result is that the VA consumption of the relay at setting is not the same on all taps, but increases slightly as the plug setting is raised. Typically, the burden of a relay nominally referred to as 3VA may vary as follows:

Plug Setting	Burden at setting
50%	2·5 VA
100%	3·1 VA
200%	3·5 VA

This effect is not of great importance, since the higher burden values correspond to the higher setting currents and therefore to low values of actual coil impedance. The actual values may be significant in the case of earth fault relays.

(b)   An induction relay is required to operate over a wide range of input current and is subject to extreme saturation at the higher values, in consequence of which the impedance of the relay varies considerably, being reduced by a factor of three (or more) with an input of twenty times the tap setting for the minimum tap (i.e. full coil).

The degree of reduction is less for higher taps, being only a factor of 1·7 - 2 for the highest tap setting, since as described above, the resistance of the winding is then a larger portion of the impedance.

8.3.3.3   Additional burden: A current transformer may be burdened by more than one device, and the sum of all such burdens must then be assessed. The transformer may also be ascribed both a 'measurement' and a 'protection' rating. For example, the same transformer may be rated 15VA Class 1·0 and 7·5VA/5P/10. This does not mean that an instrument burden of 15VA and a relay burden of 7·5VA may be applied simultaneously. The burdens are the maximum that may be applied for the respective accuracies to be obtainable. If the current transformer is to be used to energise a 3VA overcurrent relay set at 100%, and if other burdens are negligible, then, assuming a secondary winding resistance of 0·15 Ω as in the above

---

* 'Unit turns' refers to the whole coil turns.

example, an accuracy limit factor of 16 would be permissible.

Alternatively, the relay could be used in series with an ammeter consuming 2VA and with interconnecting wiring between the c.t., relay and instrument totalling 0·1 Ω (approximately 15 yd of 7/0·67mm cable). With the latter burden, the specified accuracy limit factor of 10 would be obtained.

8.3.3.4 Significance of leads: It will be noted that, in the latter example, a comparatively small amount of wiring is included as a significant burden. In fact a resistance of 0·1 Ω is equivalent to 2·5VA, which is one third of the rated output of this c.t. Although not mentioned in the former example, an allowance for wiring should always be included when dealing with c.t.s having a 5 A secondary rating. Even in the case of metal-clad switchgear with integral relay panels and very short leads, it is unwise to assume that the resistance of connections will be less than 0·05 Ω, because of the various bolted joints that will be included. Even this low figure is equivalent to 1·25VA or 16% of the rated output of the c.t. in the above examples.

Frequently, however, relays are mounted on panels located away from the switchgear. Depending upon circuit arrangements (see below) a cabling run of 15 yd may involve, with lead and return plus vertical wiring runs on the panel etc., perhaps 38 yd of circuit, with a resistance of 0·26 Ω. Making a small additional allowance for the resistance of joints as above, a total of 0·3 Ω is easily possible, which at rated current of 5 A corresponds to 7·5VA. This is the total rated output of the c.t. in the above examples without leaving anything for the useful burden (i.e. relays and instruments). A larger c.t. is therefore needed involving increased cost, and which may not always be possible (i.e. if an accuracy limit factor of 20 were required, the transformer needed might be larger than could be readily accommodated).

When the switchgear is of the e.h.v. outdoor type, the c.t. lead runs may easily be of 100-200 yards corresponding to resistances of several ohms, and burdens of useless power loss of many times the burden of the relays. If the rated secondary current is changed to a lower value, the lead loss will be reduced in proportion to the square of the change in rating, the lead resistance being unchanged. The relay burden will remain unchanged, and the available output from a c.t. of given dimensions will also be nominally the same, although some advantage may be obtained, in the case of a toroidal c.t., in the form of a lower secondary winding loss at the lower rating.

The following examples illustrate the effect :

Example 1
Current transformer rated 7·5 VA/5P/10, supplying overcurrent relay, setting 100%, 3VA burden; lead run 20 yd of 7/0·67mm cable, resistance 0·00677 Ω/yd.
Lead resistance = 40 x 0·00677 = 0·27 Ω.
Lead resistance allowing for panel connections and joints (say) 0·3 Ω.

Resistance of secondary winding of c.t.

  5A rating : 0·15 Ω, as previously
  1A rating : 2·0 Ω, see last paragraph above.

Rating	1 A	5 A	
Relay	3·0 VA	3·0	VA
Leads	0·3 VA	7·5	VA
External burden	3·3 VA	10·5	VA
External impedance	3·3 Ω	0·42	Ω
C.T. resistance	2·0 Ω	0·15	Ω
Total circuit impedance	5·3 Ω	0·57	Ω
C.T. knee point e.m.f. based on rated output	95 V	22·5	V
Accuracy Limit current	17·9 A	39·5	
Accuracy limit factor	17·9	7·9	

The above results show that the 1 A c.t. can have a smaller core than the other, since the saturation e.m.f. which is necessary to provide the rated output is less than 5 times that of the 5 A equivalent, although developed in 5 times the winding turns. The 1 A c.t. will nevertheless maintain an accurate output to the relay up to Accuracy Limit Factor of more than twice as great as will the 5 A c.t. under the stated conditions.

Example 2

Outdoor switchgear with lead run of 200 yards (7/0·67mm cable). Relay 3VA. Resistance of leads 2 x 200 x 0·00677 = 2·7 Ω.

Rating	1 A	5 A	
Relay VA	3·0 VA	3·0	VA
Relay impedance	3·0 Ω	0·12 Ω	
Leads	2·7 Ω	2·7	Ω
External burden;			
Impedance	5·7 Ω	2·82 Ω	
External VA	5·7 VA	70·5	VA
Assuming c.t. secondary			
Resistances as before	2·0 Ω	0·15 Ω	
Total secondary circuit impedance	7·7 Ω	2·97 Ω	
Secondary current for a.l.f. = 10	10 A	50	A
Secondary e.m.f. for a.l.f. = 10	77 V	148·5	V
Secondary power for a.l.f. = 10	770 VA	7425	VA

The core volumes of current transformers designed to provide the above outputs, would be in proportion to the VA outputs; both transformers are realisable, and could be accommodated within the space usually available in h.v. switchgear, but the 5 A unit would be of ten times the size and considerably greater cost.

The effect of relay saturation (see Section 8.3.3.2) has not been included in the above calculations for simplicity, but would tend to increase the disparity in outputs.

It is clear that very great advantage is obtainable by using a low secondary current rating.

It will be noted that the total lead length in the above examples has been taken as twice the lead run. Fig. 8.3.3.4A shows that, for single-phase faults, two c.t.s operate in series to energise two relays and two leads, so that each c.t. is burdened with one lead plus one relay. However, it is not unusual to form the c.t. star point in the relay panel, so that it may be earthed through an earthing link in this panel, which is convenient for testing. This involves the circuit shown in Fig. 8.3.3.4B in which two leads are included in each c.t. circuit. It is important to realise the penalty in c.t. performance that is involved by adopting this second arrangement.

The longer circuit was used in the sample calculations; the principle is not changed if the circuit of Fig. 8.3.3.4A is used, except that the lead resistance will be lower for a given lead run.

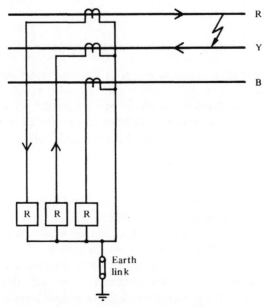

Fig. 8.3.3.4A   *Current transformers connected by 4 wires to relays*

The circuit of Fig. 8.3.3.4B will also give a poorer performance with regard to phase-fault stability; see Section 8.3.3.9 below.

8.3.3.5   Burden of earth-fault schemes: The burden of overcurrent relays usually presents little problem, but with earth-fault relays, typically with a setting range of 20-80% it is important to realise the magnitude of the burden imposed.

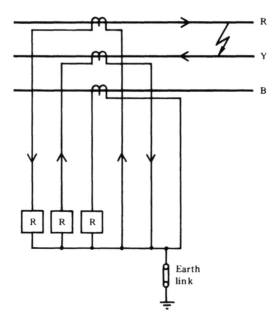

Fig. 8.3.3.4B   *Current transformers connected by 6 wires to relays*

An earth-fault relay similar in design to the overcurrent relays with a burden at setting of 3VA, if set to 20% would impose a burden equivalent to 75VA at rated current. This is the effective burden on the c.t. from the point of view of accuracy, even if rated current does not flow. With this burden, the c.t. in the former example, rated 7·5/5P/10, will become saturated if earth-fault current much above the rated value is applied.

8.3.3.6   Effective setting: The primary setting of an overcurrent relay can usually be taken to be the relay setting multiplied by the c.t. ratio. The c.t. can be assumed to maintain a sufficiently accurate ratio and, expressed as a percentage of rated current, the primary setting will equal the relay setting.

The higher burden of the earth-fault relay involves a correspondingly high exciting current component in the energising c.t. Not only this but by virtue of the residual connection, the voltage drop on the earth-fault relay is applied to the other current transformers of the paralleled group, which, in consequence, each divert an exciting current away from the residual circuit, whether they are carrying primary current or not.

The total exciting current is therefore the sum of that in the three current transformers and can be appreciable compared with the operating current of the relay; in extreme cases where the setting current is low or the current transformers are of low performance, it may even exceed the output to the relay. The effective setting current, in secondary terms, is the sum of the relay setting current and the total excitation loss. Here again, although the vector sum is strictly correct, for electro-magnetic relays it is justifiable to use the arithmetic sum, because of the similarity of power factors.

It is instructive to calculate the effective setting for a range of relay setting values. This is set out in Table 5 and the results displayed in Fig. 8.3.3.6.

Table 5

Relay plug setting		Coil voltage at setting	Exciting current		Effective setting Current	
%	A	V	I_c	$3I_c$	A	%
5	0.25	12	0.64	1.92	2.17	43.4
10	0.5	6	0.41	1.23	1.73	34.6
15	0.75	4	0.33	0.99	1.74	34.8
20	1.0	3	0.29	0.87	1.87	37.4
40	2.0	1.5	0.22	0.66	2.66	53.2
60	3.0	1.0	0.17	0.51	3.51	70.2
80	4.0	0.75	0.15	0.45	4.45	89
100	5.0	0.6	0.13	0.39	5.39	107.8

Calculation of effective setting:

Relay nominal rating 5 A
Relay burden   3 VA
(assumed to be constant)

Effective setting = relay current + c.t.. excitation loss
$$= I_R + 3I_c$$

It will be seen that, although the operating condition of the current transformers does not approach saturation, the lowest relay settings result in a large exciting loss and consequently high effective settings. The optimum relay setting is theoretically 13% but little is to be gained by applying a setting less than 20%. In fact, the lower settings are undesirable because they are unpredictable in practice.

The protection system usually has to be designed using estimated c.t. performance curves. Although, as supplied by a manufacturer, these are usually pessimistic, considerable variation is to be expected in the exciting current value taken by the actual c.t.s. When, as in the first three lines of Table 5, the effective setting is composed mainly of exciting current, the setting actually realised, and hence the relay operating time with the input fault current would be too uncertain for good grading. The effect of waveform distortion, if the fault current is high enough to drive the c.t. even moderately into saturation, would also be detrimental.

It may be thought to be desirable to apply relays having a setting range of 10-40% in all cases, the optimum setting being determined by test before com-

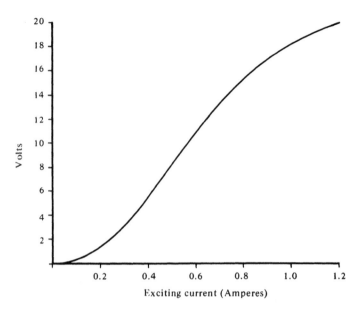

(a) C.T. excitation characteristic

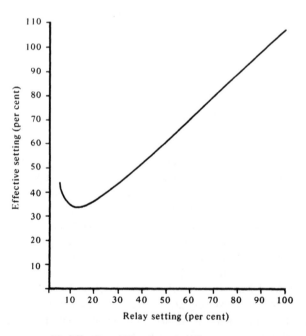

(b) Effective setting characteristic

Fig. 8.3.3.6   *Effect of c.t. excitation loss on the operation value of earth-fault relay*

missioning. It must be realised, however, that unless the earth-fault current is limited by neutral impedance, the relay may be subject to a high current, if energised by c.t.s having a high saturation e.m.f. If the setting is low, the setting multiple would then be very high and the relay would experience considerable heating.

In this respect, a tap setting of 20% is not the same when obtained from the 10-40% range as when obtained from the 20-80% range. The former range corresponds to a winding having twice the number of turns of the latter and hence a smaller wire size. When the 20% tap is selected from the 10–40% range only half the coil turns are in use, and the coil resistance is greater than when the same setting is obtained from the 20–80% range, in which case the whole coil is used.

The higher setting range is suitable for most applications and should be used unless it is known from a system study that a setting of less than 20% is necessary. When the earth-fault current is limited sufficiently by neutral earthing resistance, the lower setting range will be permissible, although frequently not necessary.

In locations where the earth resistance may be so high that greater sensitivity is needed, lower settings should be applied, and the current transformers should be specially designed to suit the higher burden and should have sufficiently low exciting currents. It is then necessary to consider the maximum earth-fault current that may flow, taking account of all possible contingencies, and to decide whether it is necessary to design the current transformers to limit the output current by saturation.

8.3.3.7   Time-grading of earth-fault relays: Earth-fault relays are graded in a similar manner to phase-fault relays. It is necessary to realise, however, that with a normal rating of current transformers, the time/primary current characteristic cannot be kept proportional to the time/relay current characteristic to the same degree that is possible with phase relays. The conditions which result in a relatively high excitation loss as discussed above are augmented still further by the fact that, at setting, the flux density in the current transformer corresponds to a point on the bottom bend of the excitation characteristic, where the exciting impedance is comparatively low. The current transformer may actually improve in performance at first with increased primary current, whereas the relay impedance decreases, until with an input current several times greater than the primary setting, the p.s.m. applied to the relay is appreciably greater than the multiple of primary current setting, which flows in the primary circuit, causing a shorter operating time than might be expected.

At still higher currents, the c.t. performance will fall off until finally the secondary current ceases to increase substantially, and waveform distortion adds further complication to any attempt to accurately assess the overall behaviour.

It is not easy to determine accurately the operating times to be expected for an earth-fault relay under such conditions without making detailed analytical calculations, but it is clear that in time grading earth-fault relays, larger margins should be allowed.

8.3.3.8   Phase-fault stability: Overcurrent and earth-fault relays are often applied together using either three or, very commonly, two overcurrent elements (see Section 8.2). When the system is subjected to phase-fault current, the earth-fault element should not be energised, but in fact, some current will flow in the residual circuit as a result of unequal exciting currents taken by the current transformers. The exciting currents may differ in part due to small differences in the c.t.s themselves and inequality of burdens and also in part due to transient conditions of the fault current.

It is unlikely with reasonable c.t.s and phase burdens that the earth-fault element will be energised to a higher p.s.m. than are the phase elements. Conditions may exist, however, where the phase elements need to be given a much larger time multiplier setting than does the earth-fault element. Such a case is shown in Fig. 8.3.3.8 in which a line feeds a delta-star transformer which supplies a lower-voltage system involving several stages of grading. The phase relays on the high-voltage side need a time setting which will grade with the lower-voltage system; no zero-sequence component can be transmitted through the transformer so that the earth-fault relay on the high-voltage side can have a low time setting.

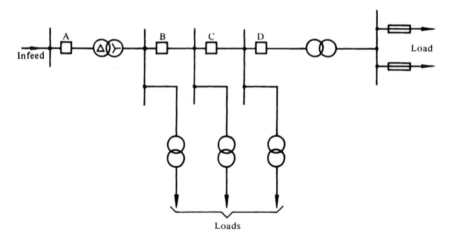

Earth fault relay at A can be set to minimum time but phase fault relays must be set to grade with following relays B, C and D.

Fig. 8.3.3.8   *System needing phase-fault stability*

In such an instance any operation of the earth-fault relay due to spill current may result in unnecessary tripping of the transformer supply before a fault on the secondary side is cleared.

Suggestions are given in Appendix B of BS 3938, 1973 (1982), 'Current transformers' (see also BS 7626, 1993) of conditions which are desirable in order to obtain phase-fault stability and accurate time grading of earth-fault relays. In particular, phase-fault stability is considered up to ten times rated current and also time grading of the earth-fault element up to ten times the earth-fault setting.

The following conditions are suggested:

(*a*)   Class 5P transformers in which the product of rated output and rated accuracy limit factor is 150 should be used.

(*b*)   The phase burden imposed on each current transformer does not exceed 50% of its rated burden.

(*c*)   The rated accuracy limit factor is not less than 10.

(*d*)   The earth-fault relay is not set below 30%.

(*e*)   The burden of the earth-fault relay at setting does not exceed 4VA.

The above requirements represent a compromise and some discretion may be used in a particular instance. For example, the use of a setting lower than that specified will give a higher degree of phase-fault stability because of the much higher impedance that the residual circuit will then have. The earth-fault current up to which good time grading can be expected will, however, be much lower.

On the other hand, the use of lower burden types, e.g. 1VA, although increasing the limiting current for time-grading, may cause deterioration of phase-fault stability. Fortunately, when combined phase-and-earth-fault relays of a low VA pattern are used, the reduction of phase-fault burden will assist the maintenance of stability and so compensate for the lower residual circuit impedance.

8.3.4   Sensitive earth-fault protection

Normal earth-fault protection is based on the assumption that fault currents that flow to earth will be determined by the earth-fault impedance of the line and associated plant. In some localities, however, the nature of the ground may be such that it is difficult to make an effective earth connection. In such a case, and particularly if continuous earth conductors are not used, a system fault to earth may result in a current which is too small to operate the normal protective system.

If a line conductor should break and fall on to semi-insulating objects, such as trees or hedges, or on to dry earth or a metalled road surface, again the fault current may be below the normal primary earth-fault setting. The circuit will, therefore, not be isolated and the conductor, which will be maintained at normal phase voltage will constitute a serious danger to life.

The difficulty of lowering the setting of normal earth-fault protection has been discussed above. The large reduction in setting from usual values (i.e. by a factor of 10 or more) which is needed to help with this problem, is not feasible with the usual type of earth-fault relay because of the very high burden which a suitably low set relay would present at rated current. The difficulty in providing current transformers to give a reasonably equivalent effective setting would then be considerable. The problem has been dealt with by the design of special earth-fault relays, based upon very sensitive elements. Both rectified moving-coil and polarised moving-armature elements have been used and electronic designs are also available. In this way the relay consumption at setting may be as low as 0·005 VA, making a setting of 1% of rated current feasible without an excessive burden on the current transformers.

It is the impedance of the burden at relay setting which is critical in estimating the effective setting. In the above example, assuming a rating of 5 A, the impedance at setting is 2 Ω. Although the current transformers will be operating on the bottom bend of their excitation characteristic, transformers of very moderate performance would not increase the effective setting to more than 50% above the nominal, i.e. to 1·5% of rating.

The relay will normally have a saturating characteristic so that the measured burden at rating will be less than proportional to that at setting. Relays of this type are of the definite-time variety. They have a back-up function, supplementary to the main protection, and are therefore given a time delay of several seconds, long enough to ensure that they do not interfere with the normal discriminative protection. Although not suitable for grading with other forms of protection, sensitive earth-fault relays may be graded as an independent system.

The limit in current sensitivity is not decided by the capability of the relays themselves but by the residual unbalanced capacitance and leakage currents that may flow in a healthy line, for which tripping is not required. This primary residual current is usually determined only by test when the relay is installed. Some range of adjustment of sensitivity is provided, and a setting is chosen which is as sensitive as possible, while remaining safe from unwanted operation.

8.3.5   High-set instantaneous overcurrent relays

The time-grading principle requires the relay of any pair which is nearer to the power source to operate more slowly. With a definite-time system, the grading increments accumulate to make the tripping time at the supply point long if several sections of grading are involved. When there is little variation in fault current throughout the feeder system, little improvement is obtained by using inverse-time relays, but when the system is long enough to cause appreciable fault current variation throughout its length, some easement is obtained. (Section 8.3.2.2, Table 2). Further reduction in overall tripping time was obtained by using extremely inverse relays and higher primary settings (Table 4).

An alternative technique consists of adding supplementary instantaneous over-current elements with high settings. It was explained in Section 8.2.2 why instantaneous relays could not in themselves form a complete system of protection.

Such relays can, however, be added to a normal graded system and set so as to operate with close up faults but to not operate with the maximum current which may flow to a fault at the remote end of the respective section as shown in Fig. 8.3.5A. A portion of the section will therefore be protected with high speed, whilst the inverse time-graded system will cover the remainder of the line.

The greatest advantage will be obtained by applying such a supplementary relay at the supply point, but some gain will be obtained by using them at other points also, provided always that there is an appreciable change in fault-current value from one end to the other of the relevant section.

A simple high-speed relay responds to the r.m.s. value, during its operating time, of the current wave; the r.m.s. value may be much increased by a transient offset component. The r.m.s. value of a wave with a maximum transient offset of typical time constant may be 50% higher than the steady state value during the first cycle. A simple fast-operating armature-type relay would respond to a current with such an offset component although the steady state value was well below the relay setting. In fact, since the first asymmetric peak current may be as much as 2·5 times the steady r.m.s. value, a high speed relay may operate with an even lower steady state current than is suggested by the above figure.

The relay is then said to overreach; that is to say that it operates for a fault further from the power source than would be calculated from steady fault current values. The percentage transient overreach is defined as:

$$\frac{I_1 - I_2}{I_2} \times 100$$

where I_1 = relay pick up current in steady-state r.m.s. A

I_2 = steady-state r.m.s. current, which when fully offset will just pick up the relay

It is important that the instantaneous relay does not operate with a fault beyond the end of the relevant section, which would invalidate the discrimination of the whole protective system, with its carefully designed grading. A setting must therefore be chosen that is much higher than the end-of-section fault current; in some cases this may result in little if any of the section being covered with certainty, i.e. in the case of no initial transient so that the fault current is not momentarily increased above the steady value.

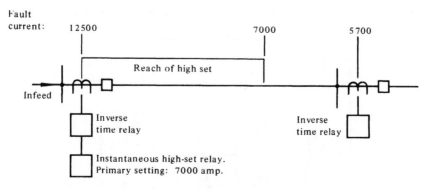

Fig. 8.3.5A   *Application of instantaneous high-set relay*

Better results are achieved by the use of relays having a designed immunity to the transient offset condition. Two arrangements of relays and auxiliary components to achieve this object are shown in Fig. 8.3.5B and C.

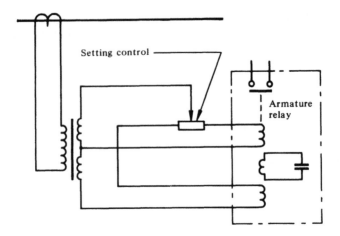

Fig. 8.3.5B   *Transient free high-set instantaneous relay*

Fig. 8.3.5B shows a relay in which the overreach is limited to about 5% for a primary system X/R ratio of 5. The electromagnet has a divided winding fed by a circuit which provides a continuously variable range of settings over a 1–4 range, e.g. 200-800% or 500-2000%.

A second winding is tuned by a capacitor. When this circuit is in full resonance, the impedance of the electromagnet is effectively increased so that the flux for a given input current is increased also. This does not occur in the first instant of energisation and moreover the amplification does not occur for the 'd.c.' component of the input current.

An alternative scheme is shown in Fig. 8.3.5C with which an even greater immunity to a transient offset is obtained by a small saturable auxiliary current transformer. This c.t. is saturated by the d.c. component of the input current to a

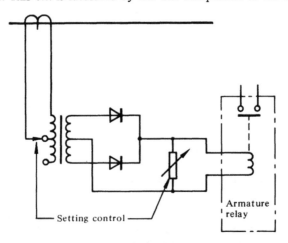

Fig. 8.3.5C   *Alternative transient free high-set instantaneous relay*

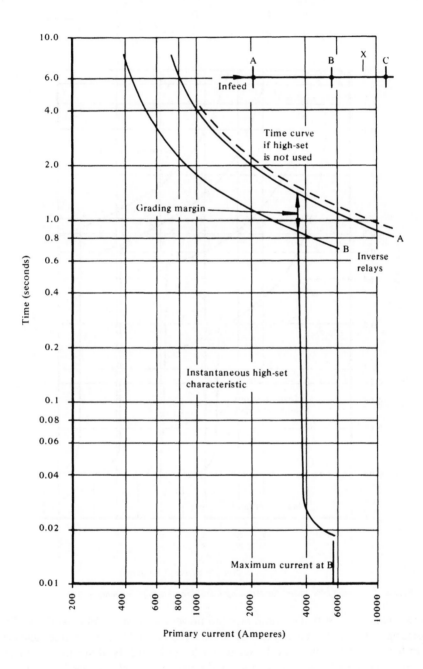

Fig. 8.3.5D   *Effect of instantanteous high-set relay on grading times*

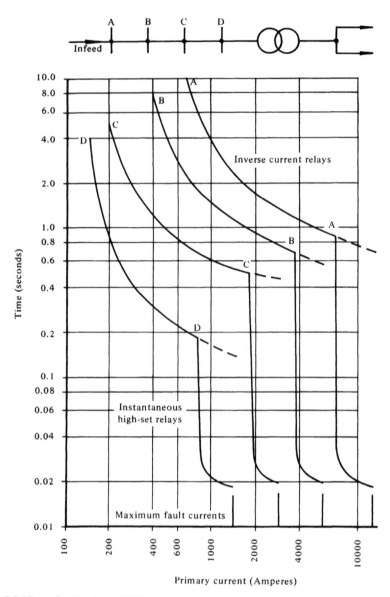

Fig. 8.3.5E *Grading for radial feeder using instantaneous high-set relays;*

degree that causes sufficient ratio error to prevent overreach almost entirely. The operating time for the same system X/R value is comparable to the former scheme.

Either of these relays may be set to a value only slightly above the maximum current calculated to flow to a fault at the end of the relevant feeder section; a good proportion of the section will then have high-speed protection provided there is a reasonable fault-current variation over the section length. This is an obvious

high-set relay

inverse relay

Fig. 8.3.5F   *IDMTL relay with instantaneous high-set element*

gain when applied to the first section from the supply point where the fault current is highest. For other sections, there is an advantage not only in clearing some faults with minimum delay, but also in the grading of the inverse-time relays for previous sections. This is illustrated in Fig. 8.3.5D, which shows grading with a high-set relay applied to the second section BC of a radial system.

The time characteristic of the first section AB must have the necessary grading margin over the high set relay time for faults at B and over the inverse-time relay for a fault at X the nominal limit of the reach of the instantaneous relay. The timing curve that would be necessary if a high set relay were not used is indicated by the broken line.

Fig. 8.3.5E shows the time grading for the radial feeder previously discussed in Section 8.3.2.2, using high-set relays.

High-set relays such as have been described above are small attracted armature elements which can be mounted in the same case as the inverse-time relay; the auxiliary components are also small and can be accommodated in the case as in Fig. 8.3.5F. No additional panel space or extra wiring is therefore needed.

8.3.6   Relay co-ordination with fuses

Primary fuses as a protective system were introduced in Section 8.2.4.1 and they are more completely discussed in Chapter 5. A brief summary here is not inappropriate.

The fault clearance time of a fuse is the sum of pre-arcing (i.e. melting) time and arcing time. With currents well above the minimum fusing value, both of these stages involve an approximately constant amount of energy, i.e. $\int i^2 dt$ is a constant. To achieve discrimination between two fuses in series it is necessary for the total $I^2 t$ taken by the smaller fuse to be less than the pre-arcing $I^2 t$ value of the larger one. This is generally the case for fuses of the same type if the ratio of ratings of two fuses in series is not less than two.

Relays may be graded with fuses provided that the relay is located nearer to the supply point than the fuse. The time-current characteristics of fuses are very steep, steeper even than that of the extremely inverse relay, and are devoid of saturation effects. Operation is very slow with currents only slightly above the minimum fusing value, but extremely fast with large currents. The short time of operation of a fuse with short-circuit current makes the use of a relay at a subsequent point impracticable.

When a relay precedes a fuse, time grading can be arranged as in the example above. It is important, while arranging the grading at the maximum current value also to ensure that the fuse characteristic does not cross that of the relay at lower current values. It was for this reason that the relay at substation D was given a primary setting as high as 125 A, although the maximum load was not expected to exceed 50 A.

8.4   Directional control

The considerations discussed in Section 8.3 have all been related to obtaining discriminative operation of relays associated with a simple radial feeder. Such a system, however, fails in the objective of maintaining uninterrupted supply to all load points, since the tripping of a circuit breaker near the feeding end of the line disconnects all subsequent substations with their loads. To remove this failing it is necessary to feed power to both ends of the line as shown in Fig. 8.4A. It is now necessary to insert circuit breakers at each end of each line section; isolation of a section can then be achieved, leaving all substations connected to one supply point or the other, and no loads are lost.

A moment's consideration will show that the circuit breakers cannot all be graded in a simple sequence as before. A given sequence must grade down in time away from the supply point, so that one grading could not be correct relative to both supply stations.

It is necessary to group circuit breakers and relays into two systems facing in opposite directions, each circuit breaker being at the start of a section, considered in the direction of its own grading system. In fact the arrangement may be regarded as the equivalent of two radial feeders fed, respectively, from either supply point, the two 'component feeders' being superimposed in the actual line (see Fig. 8.4B).

The two sets of relays are graded separately as radial feeders; the timings are indicated on Fig. 8.4B on a basis of definite-time protection with margins of 0·5 s;

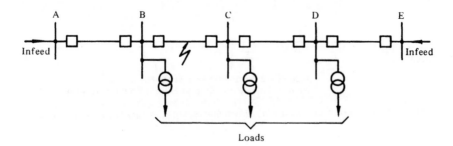

Fault is removed by tripping both circuit breakers in Section BC,
leaving all loads energised from either A or E.

Fig. 8.4A   *Distributor fed from both ends*

inverse-time relays could, of course, be applied. The complete array of time lags alone would not be satisfactory; a fault in the middle of the system would trip the breaker to the left of station B and that to the right of station D, cutting off supply from loads at B, C and D.

It is necessary that relays shall operate only with current supplied from the station which is the normal feed for the 'component feeder' to which they belong. This final requirement is achieved by giving the relays a directional response as in Fig. 8.4C.

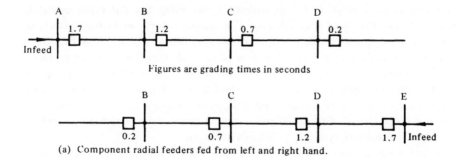

(a) Component radial feeders fed from left and right hand.

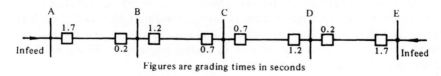

(b) Component feeders superimposed.

Fig. 8.4B   *Component radial feeders combined to form doubly fed line*

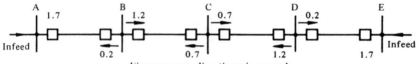

Figures are grading times in seconds.
Arrows indicate direction of current flow for tripping.

8.4C   *Doubly fed distributor showing directional response and grading times of relays*

8.4.1   Directional relays

The direction of flow of an alternating current is not an absolute quantity; it can only be measured relative to some reference which itself must be an alternating quantity. A suitable reference is the system voltage. A pure phase-comparison device would be satisfactory, but the electromagnetic relays which can achieve phase comparison are 'product' measuring devices, responsive to $IV \cos \phi$; they are, in fact, power measuring elements. The induction relay is by far the most widely used.

Torque, which is proportional to $\cos \phi$, will remain positive over a range of $\pm 90°$. A line through the phasor origin, perpendicular to the polarising voltage phasor will divide the diagram into nominal operation and nonoperation regions.

Since any relay requires a finite input of power to operate, the actual operation area is reduced by movement of the boundary line to a parallel position displaced from the original position by an amount representing the minimum operating power of the relay (see Fig. 8.4.1A). This diagram does not include effects of saturation or of spurious 'creep' torques which may distort the final boundary line.

Bearing in mind that it is the direction of the fault that is the sole requirement, the directional relay is made very sensitive. It is not given a calibrated setting, but typical relays will operate with 1–2% of rated power. (Rated power is the product of rated current and rated voltage). Since the system voltage may fall to a low value during a short circuit, directional relays should retain their directional properties down to a low voltage, typically 1–2% of rated value.

A directional inverse-time overcurrent relay comprises a directional element as above, and an inverse-time overcurrent element, the two elements being mounted together in one case; the combination is single phase, so that three such combined relays are needed to form a protection system. The operation of directional elements is made as fast as possible, the relays having a nonadjustable travel which is as small as is practicable. They are usually arranged to control the inverse-time element by means of a contact inserted in a subsidiary circuit, which must be closed to permit operation (Fig. 8.4.1B).

When, due to a change in the transmission system, a directional feature is required to be added to existing overcurrent relays, the directional elements can be obtained separately. If, however, the overcurrent relays have not been provided with the subsidiary control circuit brought out to accessible terminals, the direc-

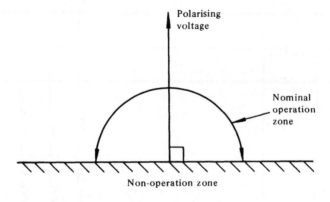

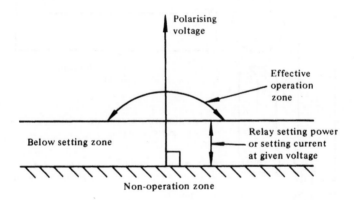

Fig. 8.4.1A   *Operation zones of directional relay*

tional relay contacts have to be connected in series with the overcurrent relay contacts in the tripping circuit. Very commonly, however, the overcurrent relay will not have separate tripping contacts for each phase brought out to separate terminals.

It is not satisfactory to connect the contacts of the three directional elements as a parallel group for insertion in the tripping circuit, since elements on unfaulted phases may operate in the reverse direction to that in the faulted phase. Instead, a polyphase relay is used, in which three electromagnets, or a polyphase electromagnet (see Chapter 6, Fig. 6.2.3F.) operate on a single moving system. The torque produced by the fault-responsive phase will then override that of the other two phases giving correct directional operation.

The above is generally true except for a power system earthed through a relatively high resistance, in which case load current may interfere with operation due to limited earth-fault current. This condition requires special study.

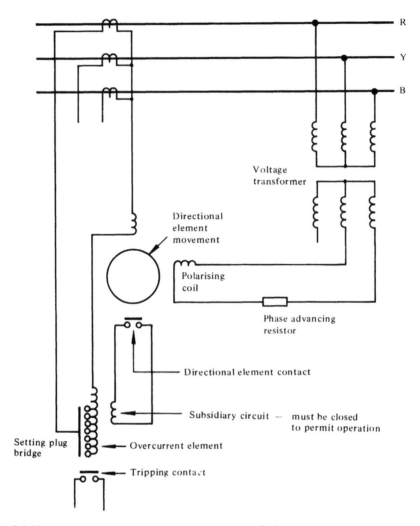

Fig. 8.4.1B   *D.. .n of directi. .l overcurrent relay — R phase*

8.4.2   Connections for directional phase-fault relays

Although wattmetric in principle, if directional relays responded to the actual fault power, their operation would not always be satisfactory. The fault power-factor is usually low, and therefore a true power measuring relay would develop a much lower torque and would be much slower than might be expected. This is, however, not the only issue.

The system voltage falls to zero at a point of short circuit. For a single-phase fault, it is the voltage across the short-circuited points which is collapsed. Referring to Fig. 8.4.2, a short circuit across Y-B phases moves the Y and B phasors together,

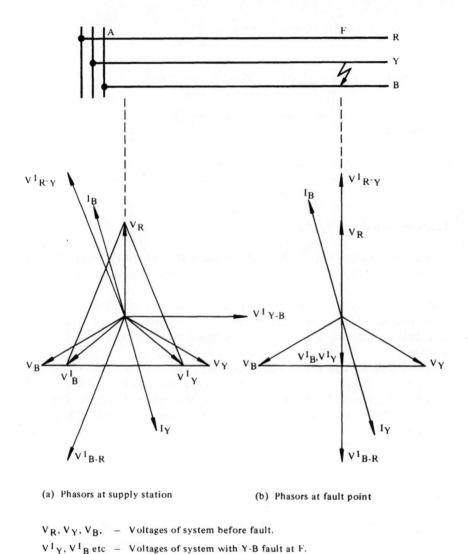

(a) Phasors at supply station   (b) Phasors at fault point

$V_R, V_Y, V_B,$ — Voltages of system before fault.

V^1_Y, V^1_B etc — Voltages of system with Y-B fault at F.

I_Y, I_B — Fault current phasors in direction A-F
                          75° lagging phase angle assumed.

Fig. 8.4.2   *Phasor diagram for Y-B short circuit*

their ends remaining on the original line Y-B. At the fault location the two phase phasors will coincide, with zero voltage across the fault and half the initial phase voltage to earth. Elsewhere the phasor displacement will be less but unbalance in magnitude and phase will exist.

The large degree of unbalance in currents and voltages would cause the torque developed by wattmetric elements connected to each phase to vary widely and even be of opposite sign as will be seen from Fig. 8.4.2.

To correct this situation, voltages are chosen to polarise each relay element which will not be reduced excessively except by a three-phase fault and which will retain a satisfactory phase relation to the current under all probable conditions.

Relay connection

The relay connection is denoted by the angle between current and voltage phasors at unity system power-factor and with the system voltages balanced.

Maximum torque angle (m.t.a.)

The maximum torque angle is the angle by which the current applied to the relay must be displaced from the voltage applied to the relay to produce the maximum torque.

The relay connection is therefore a matter of selection of suitable phases of the power system, whereas the m.t.a. refers to the inherent response of the relay itself. Both considerations need to be taken together to determine the total system behaviour.

Several different connection angles have been used with relays with varied m.t.a. Determination of the suitability of each combination involves a study of the currents and voltages of the power system and the response of each element of the relay for all fault conditions. A full analytical treatment of this subject is given by W.K. Sonnemann: 'A study of directional element connections for phase relays', *Trans.* AIEE, 1950. The results are summarised below.

8.4.2.1 $30°$ relay connection: m.t.a. = $0°$

Relay phase	Applied current	Applied voltage
R	I_R	V_{R-B}
Y	I_Y	V_{Y-R}
B	I_B	V_{B-Y}

This connection has been extensively used.

As shown in Fig. 8.4.2.1, the current leads the applied voltage by $30°$ at unity system power factor. Maximum torque is obtained when the current lags the system phase to neutral voltage by $30°$ (0.866 p.f.). Tripping zone extends from $60°$ leading to $120°$ lagging when there is no voltage distortion.

The most satisfactory m.t.a. with this connection for plain feeder protection is $0°$ (i.e. wattmetric); it can be shown that correct operation will occur for all types of fault provided three phase elements are used. When only two phase elements with an earth-fault element are used, there is a possibility of failure for one interphase condition. A Y-B fault will strongly energise a Y element with I_Y current and V_{Y-R} voltage, but the B element will receive I_B current and the collapsed V_{B-Y} voltage. The large relative phase angle of these latter quantities, see Fig. 8.4.2.1, and the low value of the voltage phasor makes the torque for this phase very poor. This is unimportant provided three phase elements are used since the Y phase will operate well, but in the case of a two-phase and one earth-fault relay, with the Y phase omitted, reliance is placed on the B phase which may fail to operate if the fault is close to the relaying point.

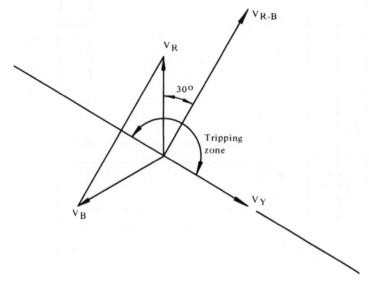

Fig. 8.4.2.1 *30° connection for directional relay*

The 30° connection is not suitable for transformer feeders. A single phase-to-phase fault on the secondary side of a star-delta transformer produces a 2-1-1 pattern of currents on the primary side, the larger current being in antiphase to the other two. This, in conjunction with the low power factor which is probable for fault current drawn through a transformer, forms a risk that one of the three-phase relays will operate in the reverse direction to the other two.

8.4.2.2 60° relay connections: m.t.a. = 0°

Two arrangements are discussed by Sonnemann, which give a 60° angle in the quantities supplied to the relay.

One that is described as 60° No. 1 connection consists of supplying the R phase relay with R-Y current and R-B voltage, and the other phase relays in a cyclic

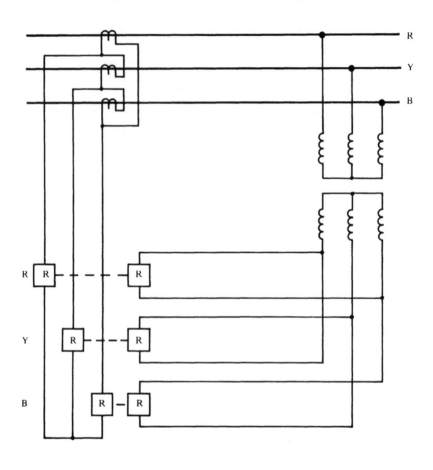

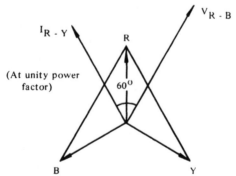

Fig. 8.4.2.2   *60° No. 1 connection for directional relay*

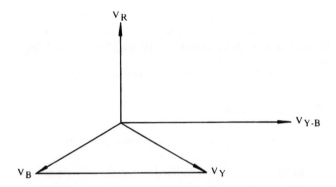

(a) Basic diagram for polarising voltage – R phase

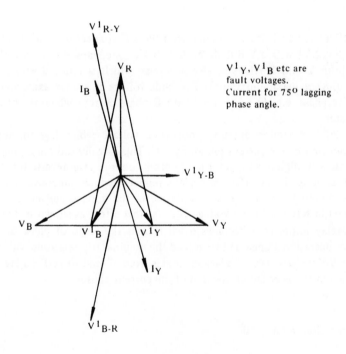

V^1_Y, V^1_B etc are fault voltages. Current for 75° lagging phase angle.

(b) Phasors for Y-B fault with voltage distortion

Fig. 8.4.2.3A *90° connection for direction relay*

manner as in Fig. 8.4.2.2. This arrangement is considered to be reliable but necessitates delta-connected current-transformers which makes their use for other functions less convenient. This connection is, therefore, rarely used. The 60° No. 2 connection consists of giving the R phase relay, R current and B to neutral voltage reversed. This connection involves risk of maloperation for all types of fault and is therefore not now used.

8.4.2.3   90° relay connection

This connection is very widely used, the relay quantities being as set out below.

Relay phase	Applied current	Applied voltage
R	I_R	V_{Y-B}
Y	I_Y	V_{B-R}
B	I_B	V_{R-Y}

The difficulty, which is encountered with the 30° connection, is avoided. Referring to Fig. 8.4.2.3A with a Y-B fault, even with V_{Y-B} collapsed to zero at the relaying point, both Y and B relays receive a voltage which is reduced very little (i.e. to (i.e. to 0.87 x maximum value). The fault voltage is always associated with the nonactive phase, e.g. V_{Y-B} is used with R phase current which is not involved in a Y-B fault.

The 90° connection applies a polarising voltage which lags the unity power factor position of the current phasor by 90°. This is really too much, since it makes operation with high power-factor fault current poor or even reverse for some conditions. It is better if the effective angle is reduced, which can conveniently be done by incorporating a phase advancing device in the voltage coil circuit. A resistor connected in series with the voltage coil has this effect as shown in Fig. 8.4.2.3B.

The relay m.t.a. now becomes α where α is the angle of phase advance. The effective operational angle is 90- α, and the nominal tripping zone will extend over 90- $\alpha \pm 90° = -\alpha$ to 180° - α lagging, that is from $\alpha°$ lead to 180°- α lag, measuring from the unity power factor position of the current phasor.

90° connection m.t.a. = 30°

The ratio of resistance to reactance in the voltage coil circuit is chosen to advance the voltage applied to the coil by 30°.

Maximum torque is produced when the input current lags by 60°, and the nominal zone of operation extends from 30° lead to 150° lag from the unity p.f. position of the current phasor.

A relay having an m.t.a. as above and quadrature connected may be used for the

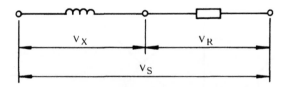

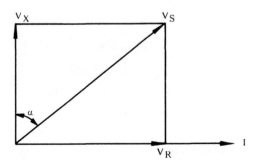

I   — Current in the coil
V_X — Voltage drop on reactance of coil
V_R — Voltage drop on resistance of
       circuit including coil resistance.
V_S — Supply voltage
a — Angle of phase advance

Fig. 8.4.2.3B   *Phase advance by series resistance in voltage coil circuit*

protection of a plain feeder with the source of zero sequence current (i.e. neutral earth point) behind the relaying point.

$90°$ connection m.t.a. = $45°$

The voltage coil circuit resistor is adjusted to give a phase advance of $45°$ giving maximum relay torque with a current lagging by $45°$ from the u.p.f. position; the zone of operation extends from $45°$ leading to $135°$ lagging. The conditions are shown in Fig. 8.4.2.3C. This arrangement, often referred to as the $45°$ connection, is suitable for all feeders. It is essential for transformer feeders or where the system earthing is such that a 2-1-1 distribution of currents is likely to occur.

    The universal suitability of the $45°$ connection, with a minimum of risk of incorrect operation, makes this the most suitable arrangement for general application.

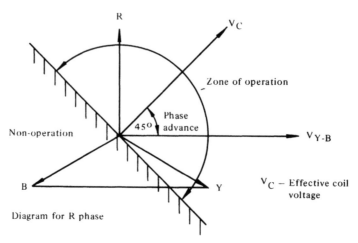

Fig. 8.4.2.3C *'45° connection' for directional relay*

8.4.3   Directional earth-fault relays

The arrangements described in the preceding Section 8.4.2 apply to phase-fault relays, although, of course, their response has to be considered for all types of fault. When for the purpose of obtaining greater sensitivity, earth-fault relays energised from the c.t. residual circuit, as described in Section 8.2, are used, a related directional element is needed.

8.4.3.1   Polarisation by residual voltage

The residual current may be derived from any phase and it is necessary to provide a polarising quantity having a related phase angle. This is found in the residual voltage of the power system.

The residual voltage is the vector sum of the three phase voltages, and may be derived from an open-delta connected residual winding of a three-phase voltage transformer, or an open-delta connection of the secondary windings of three single-phase transformers. The residual voltage is zero for balanced phase voltages, but during an earth fault on a system with a solidly earthed neutral, it is equal to the voltage depression of the faulted phase, and is in this way related to the fault current. When the system neutral is earthed through a resistance, the fault current is less and does not cause as much collapse of the phase-to-neutral voltage. The line at the point of fault is brought to earth potential and the neutral point is raised above earth potential by the voltage drop on the earthing resistance due to the fault current flowing in it. This displacement of the neutral potential is added to the voltage on the two sound phases, causing their voltage to earth to be increased. The residual voltage is the vector sum of all the phase voltages and may be quite high,

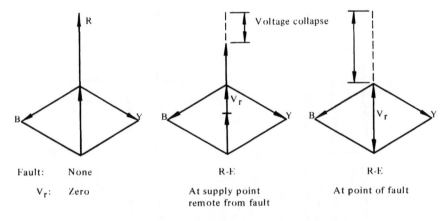

Fault: None   R-E   R-E

V_r: Zero   At supply point   At point of fault

                              remote from fault

Solidly earthed system

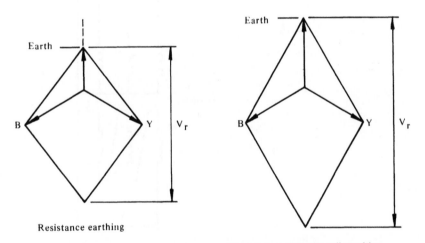

Resistance earthing

Insulated or Petersen coil earthing

Fig. 8.4.3.1A   *Residual voltage with various system conditions*

approaching three times the normal phase voltage (Fig. 8.4.3.1A).

When the system neutral is earthed through a relatively high resistance, or through an arc-suppression coil (Petersen coil) or is insulated, the neutral becomes fully displaced by an earth fault and the residual voltage is the full value of three times the normal phase voltage. These cases however, are not relevant to directional earth-fault protection, since the earth-fault current is small or zero.

When the voltage transformer is not provided with a winding for open-delta connection, a residual voltage can be extracted by a star — open delta connected group of auxiliary voltage transformers.

It is important to remember that the main v.t. primary neutral point must be

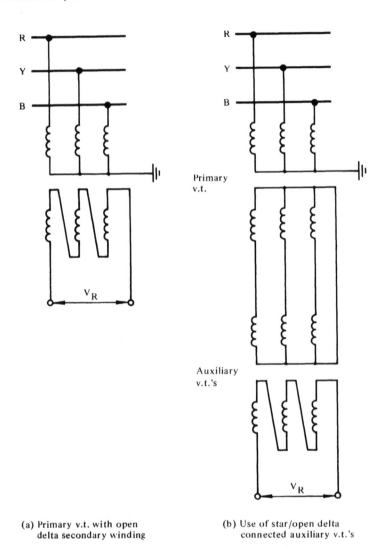

(a) Primary v.t. with open   (b) Use of star/open delta
delta secondary winding   connected auxiliary v.t.'s

Fig. 8.4.3.1B   *Arrangements for extracting residual voltage*

earthed and the secondary neutral point must be connected to the neutral point of the auxiliary group, otherwise zero-sequence currents cannot flow in the windings. The main v.t. must be suitable to transform a zero-sequence voltage and both main and auxiliary v.t.s must be rated for the rise of sound phase voltage that may occur with neutral displacement during an earth fault. Typical arrangements are shown in Fig. 8.4.3.1B.

The residual voltage is equal to three times the zero-sequence voltage drop in the source impedance; remember that the actual value of an earthing resistor is multiplied by three for inclusion in the zero-sequence network. The residual voltage is therefore displaced from the residual current by the characteristic angle of the source impedance.

When the system neutral is earthed through a resistance, this will predominate over other impedances, making the earth-fault power factor high and so the phase angle between residual current and voltage small. It will be satisfactory for the relay m.t.a. to be 0°, i.e. the relay to be of simple wattmetric construction. Any slight power-factor angle of the earth-fault current is then unimportant. It should be remembered that a deviation in angle of as much as 30° only reduces the relay torque by less than 14%. In the case of solidly earthed systems, the source impedance is reactive and may be of very high angle, even up to 87°. A further complication arises from the likelihood of positive phase error in the v.t. residual voltage output. The open-delta connection involves voltage drop in three v.t. windings so that the error is larger than would occur with star connection. The error may well make the effective angle between the secondary residual quantities exceed 90°, causing incorrect relay operation.

Relays for this duty have not always been provided with phase-angle compensation, but realisation of these facts some years ago promoted the opinion that where some compensation could be provided at little cost, this should be done.

Consequently, a preference was stated in a relay specification for angle compensation of 14°, which amount could be produced by a shading loop fitted on the voltage electromagnet of typical directional elements.

It must be realised that this m.t.a. of 14° is not designed to provide optimum torque, nor is it related to resistance earthed systems for which a correction of 14°, even if theoretically optimum, would be insignificant. The sole purpose is to ensure that reverse operation is not caused by excessive secondary phase angle arising as described above, for a solidly earthed power system.

Alternatively, a larger amount of angle compensation can be provided by a capacitor and resistor connected in series with the relay voltage coil (see Fig. 8.4.3.1C).

Exact compensation is not necessary, provided the current phasor falls well within the operative zone. As mentioned above, loss of sensitivity and decrease of speed are not serious for an angle between final operating phasors of at least 30°. Compensation is, therefore, usually adjusted to give a maximum torque angle of either 45° or 60°.

8.4.3.2 Polarisation by neutral current

If the residual voltage may be too low, or if voltage transformers are not available, or not suitable to supply residual voltage, the neutral current of a power transformer or of an earthing transformer can be used to polarise directional relays.

The relay must have a low-impedance 'current' winding in place of the more usual 'voltage' winding. Moreover since the neutral current and residual current from the line c.t.s are in phase the relay must be designed to operate for this condi-

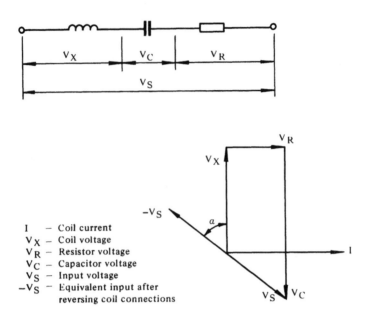

I   — Coil current
V_X — Coil voltage
V_R — Resistor voltage
V_C — Capacitor voltage
V_S — Input voltage
$-V_S$ — Equivalent input after
reversing coil connections

N.B. The compensator produces a lagging phase shift of more than 90^0; with fault current lagging V_S, relay torque would be reverse to that without the compensator; this is corrected by reversing the coil connections, which is equivalent to reversing V_S.

Fig. 8.4.3.1C   *Phase-angle compensation for earth-fault directional relay*

tion; this involves a phase shift being introduced, usually by shunting the more reactive electromagnet of the relay with a resistor.

The neutral current of a star-delta transformer will always flow from the earth into the system; residual current flows towards the fault and may pass through a group of line c.t.s in either direction. The relay can therefore discriminate between faults in the alternative directions.

When more than one transformer, each with an earthed neutral, is operated in parallel in one station, current transformers must be installed in each neutral and connected in parallel to supply the relay; either power transformer can then be switched out without invalidating the protection.

Special cases arise of star-star transformers with both neutrals earthed or of autotransformers with an earthed neutral. These cases require special study; only if the neutral current always flows towards the system, regardless of the fault position, can current polarisation be used.

8.4.3.3   Dual polarisation

In extreme conditions, the residual voltage may be low in a station because of the large rating and therefore low impedance of the installed transformers, while at times there is a possibility of all transformers being disconnected. In the latter circumstance, the residual voltage during a system earth fault would be higher.

Relays are available which can be simultaneously polarised by neutral current and residual voltage; the two quantities do not mutually interfere but both react with the line residual current to produce active relay torque. Correct operation is thereby assured for all system conditions.

8.4.4 Grading of ring mains

The need for a directional feature was introduced in Section 8.4 by considering a transmission line interconnecting two main-supply stations. This rather limited example may now be extended to the general case of the ring main shown in Fig. 8.4.4, which shows a main-supply station A connected via a ring feeder to substations B, C, D and E. Circuit breakers are installed at each end of each line

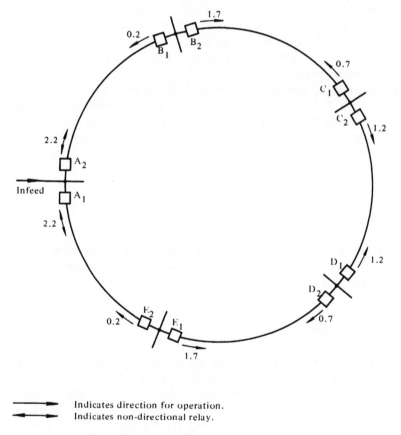

	Indicates direction for operation.
	Indicates non-directional relay.

Fig. 8.4.4 *Grading scheme for ring main*

section; as before the system can be regarded as two superimposed radial feeders.

Directional overcurrent relays are installed and must be graded each way round the ring in two sequences of relays which 'look' in the alternative directions. Thus if the circuit breakers on either side of the various stations are denoted A_1, A_2 : B_1, B_2 etc. the two tripping sequences are B_1-C_1-D_1-E_1-A_1 and E_2-D_2-C_2-B_2-A_2.

The nominal grading times of 0·5 s intervals are the usual values of 0·2 – 0·7 – 1·2 – 1·7 – 2·2, as indicated in Fig. 8.4.4. The arrows on the diagram indicate the direction of current flow for operation of each relay. At the supply point A, power can flow only away from this station and so directional discrimination is not required. A phase-fault directional element would in any case tend to have its contacts permanently closed by load power. There is little purpose, therefore, in installing other than nondirectional relays at A_1 and A_2.

At other points in the ring, the two relays facing in opposite directions may have different timings. Where definite time grading is used the difference will be one or more grading margins except that, with an even number of feeder sections, the two relays at one station will have the same setting. Where there is a satisfactory margin, the slower of the two relays need not be made directional.

When inverse-time relays are used, the timings may be similar to those shown in Table 2, Section 8.3.2.2. The timings need then to be reassessed at the maximum current value appropriate to the faster relay before any decision to omit a directional element is made.

The practice of omitting some directional elements, at other points than the supply station A, is not recommended except for very small rings. The operating time of the directional element is not zero, although it is not specifically allowed for in a grading calculation. When directional and nondirectional relays are mixed, some extra margin should be allowed in the calculated grading times, to cover the operation of the directional elements.

8.4.5   Multiple-fed ring mains

Time-graded overcurrent protection cannot be applied to ring mains with two or more power sources. In such cases it is necessary to apply distance protection or some form of unit protection.

8.4.6   Parallel feeders

A pair of parallel feeders is the limiting case of a ring main. In Fig. 8.4.6 directional relays are installed at the receiving ends B and D, of two parallel feeders, and nondirectional relays can be used at the feeding ends A and C. Relay B must grade in time with relay C, and relay D with relay A. In practice, relays B and D are given low settings, typically 50% current setting and 0·1 t.m.s. Care must be taken to ensure that the continuous rating of twice setting current is not exceeded, since load current will be carried without causing operation.

Relays A and C are set to provide the usual grading margin with relays B and D.

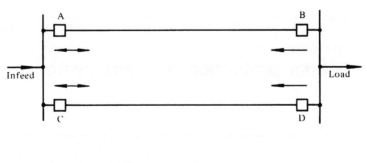

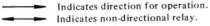

Indicates direction for operation.
Indicates non-directional relay.

Fig. 8.4.6   *Protection of parallel feeders*

8.5. Bibliography

Books

Protective relays: their theory and practice by A R Van C Warrington (Chapman & Hall, 3rd Ed., 1977)
Protective relays application guide (GEC Measurements, 1975)

Papers

'A study of directional element connections for phase relays' by W K Sonnemann (*Trans.* AIEE 1950, **69**)
'Protective relaying on industrial power systems' by C W Lathrop and C E Schleckser (*ibid.*, 1951, **70**)
'Selecting a.c. overcurrent protective devices for industrial plants' by F P Brightman (*ibid.*, 1952, **71**)
'Overcurrent relay co-ordination for double-ended industrial substations' by G R Horcher (*IEEE Trans. Ind. Appl.,* Nov/Dec 1978, **IA-14**, 6)

Chapter 9

Feeder protection: distance systems

by L. Jackson

9.1 Introduction

The principle of impedance measurement and the engineering of basic impedance measuring elements into schemes of distance protection whether they be switched, nonswitched, carrier-aided etc., has always formed a very important part of power-system feeder protection. The importance of its role is reflected in the fact that this form of protection more than any other has long been the subject of rapid and continuous development necessary to meet the ever increasing demands typical of modern power systems. Such development has, for many years, seen little change in operating principle (discounting more recent computer-based techniques) although the methods of implementation vary considerably largely because of the degrees of freedom allowed by modern developments in technologies of all kinds (component, computer, communications etc.). However, although implementation of the principle is subject to continuous change through development, there are basic design concepts which are almost universally established and which are important in understanding the principles of application. For this reason, this chapter is concerned mainly with the theoretical basis for design of distance protection and the more fundamental aspects of application.

9.2 Historical

Distance relays were introduced in 1923 in an effort to overcome the application problems associated with graded time overcurrent relaying. The first relays were based on the induction-disc principle which was designed specially to have a linear time-distance (or impedance) characteristic in which the speed of operation was proportional to the fault loop impedance, i.e. $t \propto Z$. Such relays located at various points in a system would tend to operate in such a way that the relay closest to the fault operated first. When used in association with directional elements the scheme provided a highly discriminative form of protection without the need for a communication link and also provided a back-up protection.

In 1931, Lewis and Tippet developed the classical theory of distance relaying which provided the stimulus for rapid development in this field of protection. Out of this development came the concept of stepped time-distance relaying which was to overcome the problems associated with time-dependent impedance measurement in developing transmission systems. Stepped time-distance relaying is still the basis for modern distance protection, as will be shown later, in which the object is to ensure a minimum operating time for a given impedance range, the time being increased for higher impedance ranges.

Also during this period of development, the understanding of the requirements of distance relaying increased rapidly and new characteristics emerged (such as the mho relay and reactance relay) which were better suited to the requirements of power systems.

In more recent years, the establishment of electronic methods has provided a much greater design flexibility which is reflected mainly in the development of new types of comparator having more complex characteristics although it should be noted that such flexibility has not always been developed to the best advantage, particularly in the early years of development.

9.3 Operating principles

9.3.1 Impedance measurement

The operating principle of distance protection is based on the fact that, from any measuring point in a power system the line impedance to a fault in that system can be determined by measuring the voltage and current at the measuring point. In practice, the measuring relay is arranged to have a balance point such that, at a particular fault setting, a current or voltage proportional to the primary fault current is balanced against a current or voltage proportional to the primary voltage. This balance point is defined by the relay impedance setting, Z_R. Thus, the relay either operates or restrains, depending on whether the fault impedance is less than, or greater than, the relay setting. The setting of the relay must be matched to the primary system and because the latter comprises complex impedances it is necessary to take account of the phase angle of the protected line. This is considered in more detail later.

Relays such as described above can be used at various points on the system and by arranging the relays so that those nearer to the fault operate sooner than those more remote, discriminative tripping of the circuit breakers controlling the various feeders can be achieved. Such discrimination requires, in addition to impedance measurement, a directional feature and a time-dependent feature. The latter is obtained by providing a combination of distance relays with a stepped time-distance characteristic as shown in Fig. 9.3.1A. In the particular example shown, the characteristic of the distance protection at any of the locations A, B, C or D is

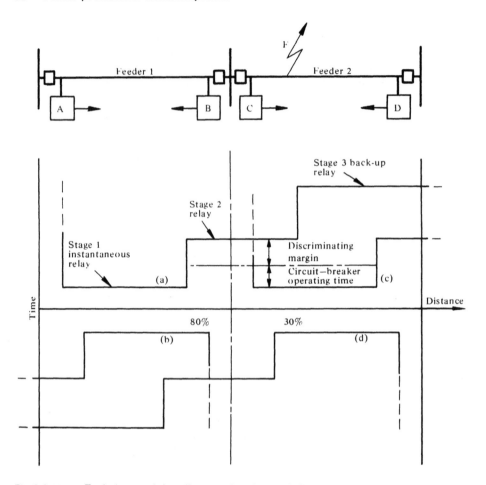

Fig. 9.3.1A   *Typical stepped time-distance relay characteristic*

in 3 stages. Each of the relays has a directional impedance measuring facility.

The relay at A has a stage 1 impedance range which is set to operate (instantaneously) and trip the associated breaker when a fault occurs within the first 80% of feeder 1 and only when the fault current is in the direction shown. The stage 2 setting of the same relay is set to include the whole of feeder 1 and (usually) the first 20-30% of feeder 2 but tripping is initiated only after a short time delay. The stage 3 setting is set to cover faults in feeders 1 and 2 and possibly beyond with tripping taking place after a further time delay. Combining these stages of measurement and superimposing those of relays B, C and D relative to the two feeders and with direction of measurement indicated by the arrows, results in the overall time-distance characteristic shown in Fig. 9.3.1A. It is clear that any fault on feeder 2 is isolated instantaneously by operation of the relay at C and because the relay at A operates only after a time-lag for such faults, feeder 1 remains in service.

The settings quoted above are typical and in the case of stage 1 the object of setting only to 80% is to allow sufficient margin for relay and system impedance errors and thus avoid incorrect tripping for faults on the adjacent feeder. Stage 2 is set well into the adjacent feeder to ensure definite operation for faults not covered by stage 1. The stage 3 provides a number of features, depending on the overall scheme arrangement, which are described later.

9.3.2   Derivation of basic measuring quantities

In addition to the basic requirements outlined above, it is necessary for a distance protection to accommodate the multiplicity of power system faults (three-phase, phase-to-phase, phase-to-earth, two-phase-to-earth and three-phase-to-earth) by arranging that the appropriate relays measure the same impedance for all fault types. This is achieved by deriving the correct voltage and current quantities from the power system appropriate to the type of fault condition. Thus, for faults involving more than one phase the appropriate phase-to-phase voltage is applied to the relay together with the difference of the phase currents as illustrated in Fig. 9.3.2A. By this means the measured impedance for phase-to-phase and three-phase faults is the positive sequence impedance Z and it can be shown that this impedance is also measured for a double-earth fault condition. It should be noted that impedance measurement based on phase-to-phase voltage and phase-current (rather than the difference of phase currents) results in different impedances for all three types of multiphase fault.

Measurement for a single-phase-earth fault condition is somewhat more complicated even for the simple system illustrated in Fig. 9.3.2A. To use phase-to-phase voltage and the difference in phase current, as in the case of phase-fault measurement, results in a complex impedance which is not only in excess of the positive sequence impedance of the line but also varies with zero and positive sequence source impedance. Measurement of positive sequence impedance is assured only by applying phase-to-neutral voltage to the relay together with a combination of line and residual currents. Thus, for a single-phase-earth fault the voltage at the relaying point is:

$$V = I_1 Z_1 + I_0 Z_0 + I_2 Z_2$$

Putting $Z_1 = Z_2$,

$$V = Z_1 (I_1 + I_2 + I_0) + I_0 Z_0 - I_0 Z_1$$

$$= IZ_1 + I_0 \left(\frac{Z_0}{Z_1} - 1\right) Z_1$$

where $I = I_1 + I_2 + I_0$

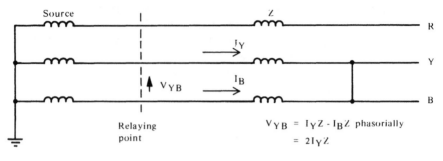

$$V_{YB} = I_Y Z - I_B Z \text{ phasorially}$$
$$= 2 I_Y Z$$

$$\frac{V_{YB}}{I_Y} = 2Z \qquad \frac{V_{YB}}{I_Y - I_B} = Z$$

(a) Phase-to-phase fault

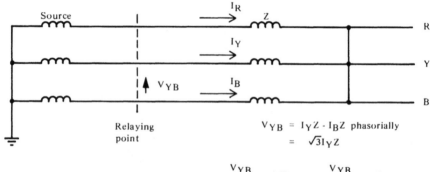

$$V_{YB} = I_Y Z - I_B Z \text{ phasorially}$$
$$= \sqrt{3} I_Y Z$$

$$\frac{V_{YB}}{I_Y} = \sqrt{3} Z \qquad \frac{V_{YB}}{I_Y - I_B} = Z$$

(b) Three-phase fault

Fig. 9.3.2A   *Phase-fault measurements*

Thus

$$Z_1 = \frac{V}{I + I_r \dfrac{1}{3}\left(\dfrac{Z_0}{Z_1} - 1\right)}$$

where $I_0 = \dfrac{I_r}{3}$

i.e the current applied to the relay is the appropriate phase current I and a residual current component

$$\frac{I_r}{3}\left(\frac{Z_0}{Z_1} - 1\right)$$

The above analysis is inadequate for power systems which have multiple earthing where a fault to earth results invariably in fault current flowing in the healthy phases, which consequently results in impedance measurement errors. The problem is illustrated by Fig. 9.3.2B which shows a simple system grounded at both ends and with a single-phase-ground fault; a delta/earthed-star transformer at one end of

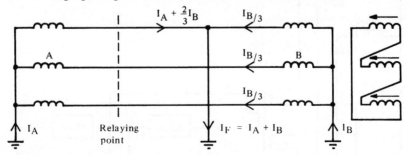

Fig. 9.3.2B *Multiple earthing and sound-phase currents*

the system can generate zero-sequence current. The sound phase currents shown can have the effect of reducing the apparent impedance to the fault which effectively extends the stage-1 zone of the protection resulting in overreach and possible loss of discrimination. The method of earth-fault compensation to correct this condition is shown in Fig. 9.3.2C and is only one of many which may be used based on the injection of a residual current component into the relay terminals. It can be shown that the proportion of residual current required to ensure measurement of positive sequence impedance is given by

$$K = \frac{1}{3}\left(\frac{Z_0}{Z_1} - 1\right),$$

i.e. the same factor as derived earlier for the simple single-end fed condition.

The above describes how different relaying quantities are necessary to distinguish between phase and earth faults, common practice being to provide two separate sets of (stage-1) relays for these faults. Each set comprises three relays because phase faults can involve any pair of phases and similarly any individual phase can be faulted to earth. Stage 2 impedance measurement may use stage-1 relays, the setting of which are increased after a stage-2 time-lag (zone-switched), or may use six separate relays. Stage-3 measurement uses six separate relays one of the functions of which may be to initiate the zone-switching where it is used.

9.4 Impedance-measuring elements (comparators) and their characteristics

9.4.1 Presentation of characteristics

The principles outlined above have shown that a distance relay is designed to have a

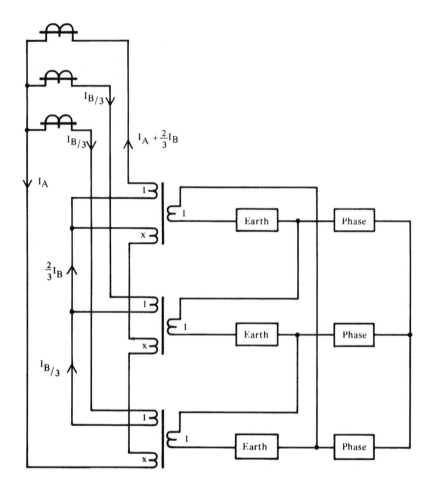

Fig. 9.3.2C   *Earth-fault compensation*

balance point so that operation occurs for line impedances less than the setting value of the relay. It is important, however, that the characteristic of the relay is not confined only to operation along the feeder length but should cater also for the effects of extraneous resistance at the point of fault while retaining adequate immunity against extreme system conditions such as high circuit loading, power swings etc. There are numerous types of characteristics possible, particularly now that there is much greater design freedom using electronic techniques, and it is convenient therefore to have a means of comparing the various types. Such means is provided by the polar characteristic. This form of presentation describes the locus of the relay operating point as the angle between the line voltage and current is varied and is usually referred to a phasor defining the impedance of the protected line.

The measuring element itself may be a device which compares magnitude of

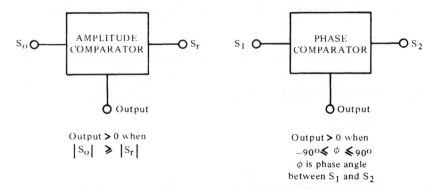

Fig. 9.4.1A   *Block representation of ideal comparators*

input quantities (amplitude comparator) or a device which compares the phase angle between inputs (phase comparator) as shown schematically in Fig. 9.4.1A, together with the respective operating criteria. Practical comparators have until recent years been confined largely to the two-input arrangements shown. Now, however, there is increasing use of multi-input comparators, as well as digital techniques, designed to produce more complex polar characteristics.

The manner in which various types of comparator operate will be considered in more detail later, it only remains to note that the quantities S_o and S_r or S_1 and S_2 in Fig.. 9.4.1A, are functions of the primary system current and voltage. In most practical cases, some form of phasor mixing is required to yield the various relay characteristics to be described.

9.4.2   Derivation of relay characteristics

As stated, the basic measuring element may be of the amplitude type as depicted in Fig. 9.4.1A. In general, operation of the amplitude comparator is defined in terms of the signal S_o applied to the operate input and the signal S_r applied to the restraint input such that, for operation,

$$|S_o| \geqslant |S_r|$$

Usually, S_o and S_r are derived by the phasor mixing of signals proportional to the power system current and voltage such that:

$$S_o = K_A I + K_B V$$
$$S_r = K_C I + K_D V$$

where I and V are r.m.s. phasor values of current and voltage.

Thus, by suitable choice of the constants K_A to K_D, which may be complex, it is possible to obtain a family of characteristics. The most basic characteristic is the plain impedance which is defined when $K_B = K_C = 0$, $K_A = Z_R$ and $K_D = 1$, where Z_R is complex, the argument being approximately equal to that of the line so as to ensure correct orientation of the polar characteristic in relation to the line phasor.

Thus, operation occurs when

$$|S_o| \geqslant |S_r|$$

or

$$|IZ_R| \geqslant |V|$$

i.e. $|Z_R| \geqslant |Z|$ where $Z = \dfrac{V}{I}$

It is seen that the relay responds only to the magnitude of the impedance as defined by the applied voltage and current and is quite independent of the phase angle between these quantities. Thus the polar impedance characteristic is a circle centred on the origin as shown in Fig. 9.4.2A in which impedances within the circle correspond to relay operation.

In general, the solution of the above input equations yields a circular characteristic in which the centre and radius are controlled by the constants K_A to K_D. By suitable choice of these, the radius can be made infinite so that the circle becomes a straight line (ohm or directional) as explained in Section 9.4.4.

The impedance, defined above as Z_R, is an important component of any

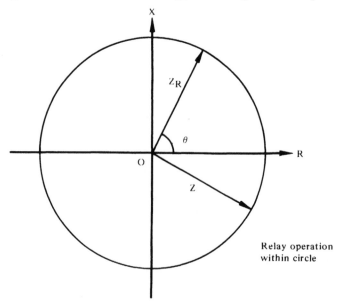

Fig. 9.4.2A   *Plain impedance polar characteristic*

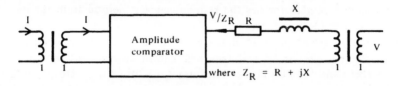

(a) Series LR impedance in voltage circuit

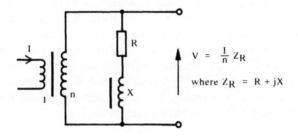

(b) Series LR impedance in current circuit

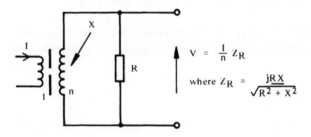

(c) Parallel LR impedance in current circuit (transactor)

Fig. 9.4.2B *Methods of providing characteristic impedance*

distance relay and is known as the characteristic impedance, sometimes referred to as the mimic or replica impedance. The main purpose of this impedance is, as stated above, to ensure correct orientation of the impedance measuring characteristic in relation to the impedance of the protected circuit. For most relay designs, it is usually arranged that the best performance, particularly in terms of operating speed, is achieved at the characteristic angle, θ, so that, in practice, the characteristic impedance is aligned with the impedance vector of the protected circuit to ensure optimum high-speed clearance for solid faults.

Inevitably, the phase-shifting property of the characteristic impedance will affect the dynamic performance of the distance relay so that the type and location of the impedance is important. In establishing the impedance to be used with analogue forms of protection it is assumed for simplicity that the impedance of the primary circuit is a series LR circuit. Thus, one arrangement of characteristic impedance, which has been used in practice, is as shown in Fig. 9.4.2B(*a*) in which a current operated comparator has the restraint current derived from the restraint voltage via a series LR impedance.

An alternative arrangement which has come to be preferred uses a mimic impedance in the current circuit and two well known arrangements are shown in Figs. 9.4.2B(b) and 9.4.2B(c). The transformer reactor or transactor is, in fact, a parallel LR equivalent and as such is only a true electrical mimic of the primary circuit impedance under steady-state conditions. Under dynamic conditions, the transactor acts as a filter to d.c. offset transients, a feature which results in improved dynamic measuring accuracy when used with the appropriate comparator. Other practical advantages of the transactor include the fact that only one iron-cored element is needed for which the transient flux levels and therefore core size is less than that required in the auxiliary current transformer of Fig. 9.4.2B(b). A practical transactor design invariably includes means for adjusting the phase angle and relay impedance settings usually by tapped primary and secondary windings and variable resistors across the secondary output winding.

9.4.3   Equivalence of amplitude and phase comparators

The operating boundary of the characteristics considered above correspond to marginal operation of the amplitude comparator, a condition which is defined by the equation

$$|S_o| = |S_r|$$

For such a condition it can be seen from Fig. 9.4.3A(a) that the sum and difference phasors (S_1 and S_2) of phasors S_o and S_r are always at right angles. Thus, the amplitude criterion of the simple amplitude comparator can be equated to a phase criterion of $\pm 90°$ for a phase comparator provided the input to the latter is derived as follows,

$$S_1 = S_o - S_r \tag{1}$$
$$S_2 = S_o + S_r \tag{2}$$

It is apparent from Fig. 9.4.3A(b) that when S_o exceeds S_r, corresponding to decisive operation of the amplitude comparator, the angle ϕ between S_1 and S_2 is less than $\pm 90°$. Similarly, when S_o is less than S_r as in Fig. 9.4.3A(c), corresponding to decisive restraint, the angle ϕ is greater than $\pm 90°$. Thus the amplitude criterion for operation $|S_o| > |S_r|$ can be equated to the phase criterion,

$$- 90° < \phi < 90°$$

for a phase comparator fed with signals S_1 and S_2.

It is apparent also from Fig. 9.4.3A(a) that maintaining S_o equal to S_r (corres-

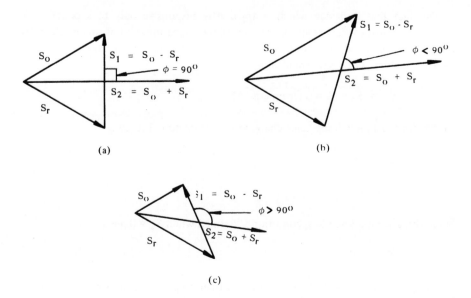

Fig. 9.4.3A   *Equivalence of amplitude and phase comparators*

ponding to a phase criterion of 90°) and allowing S_o and S_r to move more into phase has the effect of reducing S_1. In the limit when S_o and S_r are cophasal, S_1 reduces to zero, and this represents a point on the boundary of the polar characteristic. Similarly, when S_o and S_r are antiphase then S_2 reduces to zero and this represents a second point on the characteristic boundary.

Any characteristic obtained from the amplitude comparator may be derived also from the phase comparator provided the above input relationships are observed. For example, the impedance characteristic of an amplitude comparator requires that

$$S_o = IZ_R$$
$$S_r = V$$

Such a characteristic using a phase comparator from eqns, 1 and 2, required that

$$S_1 = IZ_R - V \qquad (3)$$
$$S_2 = IZ_R + V \qquad (4)$$

and that the criterion for operation is:

$$- 90° < \phi < 90°$$

where ϕ is the angle between S_1 and S_2.

The fact that a device which is apparently responsive only to a phase angle between quantities can result in an impedance measurement is made clear by considering marginal operation when $S_1 = 0$ and S_o and S_r are cophasal as explained above. For such a condition

$$S_1 = S_o - S_r = 0 \text{ (from eqn. 1)}$$

so that for the simple impedance characteristic from eqn. 3 it gives

$$IZ_R = V$$

or

$$Z = Z_R$$

Similarly, a second boundary condition is defined when $S_2 = 0$, i.e.

$$S_o = - S_r$$

This concept of marginal phase comparator operating conditions is of some benefit in the next section.

The above principle relating to duality of phase and amplitude comparators is very important and has been shown to exist not only under steady-state conditions, but also under transient conditions provided the comparator systems have similar electrical force laws. In terms of more recent developments, particularly those based on electronic methods, the phase comparator has come to be favoured for reasons to be seen later. As with the amplitude comparator, the input quantities S_1 and S_2 can be written in general form as

$$S_1 = K_1 I + K_2 V$$
$$S_2 = K_3 I + K_4 V$$

where again the choice of constants K_1 to K_4 controls the relay characteristic. Constants K_1 and K_2 are usually made equal to Z_R and -1, respectively, and as such they provide the main impedance measuring characteristic of the relay.

However, a more usual form for the generalised input arrangement is:

$$S_1 = K_1 I Z_{R1} - K_2 V \tag{5}$$
$$S_2 = K_3 I Z_{R2} + K_4 V \tag{6}$$

where constants K_1 to K_4 usually do not introduce phase shifts. Impedance Z_{R1} is assumed to have an argument θ_1, and Z_{R2} an argument θ_2.

9.4.4   Basic range of impedance characteristics

The *plain impedance characteristic* derived in Section 9.4.2 and shown in

Fig. 9.4.2A represents the simplest form of characteristic and was indeed the first to be applied in the early graded-time schemes. In Section 9.4.3 it was shown how such a characteristic can be derived using a phase comparator and, particularly, how the two main measuring points on the characteristic boundary can be derived, i.e. when $S_1 = 0$, $IZ_R = V$ corresponding to $Z = Z_R$ and when $S_2 = 0$, $Z = -Z_R$. In general, a line joining the points corresponding to $S_1 = 0$ and $S_2 = 0$, assuming an operating criterion of $\pm 90°$, will be a chord of the circular characteristic and for most applications is in fact the diameter.

A *general characteristic* can be drawn from the generalised phase-comparator input arrangement of $S_1 = K_1 \, IZ_{R1} - K_2 V$ and $S_2 = K_3 \, IZ_{R2} + K_4 V$ and dividing throughout by I gives $S_1 = K_1 Z_{R1} - K_2 Z$ and $S_2 = K_3 Z_{R2} + K_4 Z$.

$$\text{For } S_1 = 0, Z = \frac{K_1 Z_{R1}}{K_2}$$

$$\text{For } S_2 = 0, Z = \frac{-K_3 Z_{R2}}{K_4}$$

These measuring points are shown in Fig. 9.4.4A and because K_2 and K_4 are real it can be seen that these points locate the diameter of the circular characteristic. Fig. 9.4.2A is a special case where

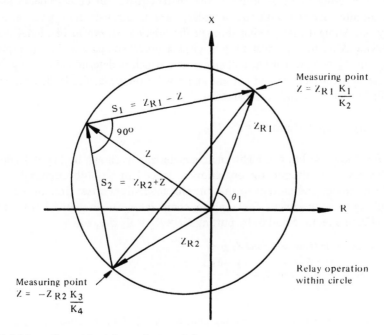

Fig. 9.4.4A   *General impedance characteristic*

$$\frac{K_1 Z_{R1}}{K_2} = Z_R \text{ and} \frac{K_3 Z_{R2}}{K_4} = -Z_R$$

but because this is a circle centred on the origin it has the obvious disadvantage of allowing incorrect operation for reverse faults within the relay setting Z_R. This problem can be overcome by using a separate directional relay with the plain impedance relay.

The *directional relay*, although strictly not an impedance measuring element, belongs to the same family as can be seen by considering the generalised phase comparator inputs (eqns. 5 and 6).

Assuming $K_1 = K_4 = 1$, and $Z_{R1} = Z_{R2} = Z_R$,

then when $S_1 = 0, \ Z = \dfrac{Z_R}{K_2}$

and if $K_2 \longrightarrow 0, \ Z \longrightarrow \infty$

when $S_2 = 0, \ Z = -K_3 Z_R$

which for $K_3 \longrightarrow 0, \ Z \longrightarrow 0$

With one measuring point at infinity, the characteristic can be considered as a circle of infinite diameter with the boundary passing through the origin, i.e. the boundary is a straight line passing through the origin as shown in Fig. 9.4.4B(a). Thus, making K_2 and K_3 equal to zero gives a phase comparator having inputs $S_1 = IZ_R$ and $S_2 = V$, which has a characteristic which is determined only by the phase-angle between voltage and current and not by their magnitude. The equivalent inputs* for an amplitude comparator are:

$$S_o = IZ_R + V \text{ and } S_r = V - IZ_R$$

The *ohm relay* also has a straight line characteristic as shown by Fig. 9.4.4B(b) but in this case the criterion for operation is dependent on both magnitude and phase of the voltage and current so that the characteristic is offset from the origin. The input signal arrangement for a phase comparator can be derived from the generalised inputs (eqns. 5 and 6) by putting $K_1 = K_2 = K_3 = 1$, and

* Defining S_o and S_r in terms of S_1 and S_2 gives

$$S_o = \frac{S_1 + S_2}{2} \text{ and } S_r = \frac{S_2 - S_1}{2}$$

Usually the factor of ½ in both S_o and S_r is omitted since it does not affect the operating criterion of the comparator.

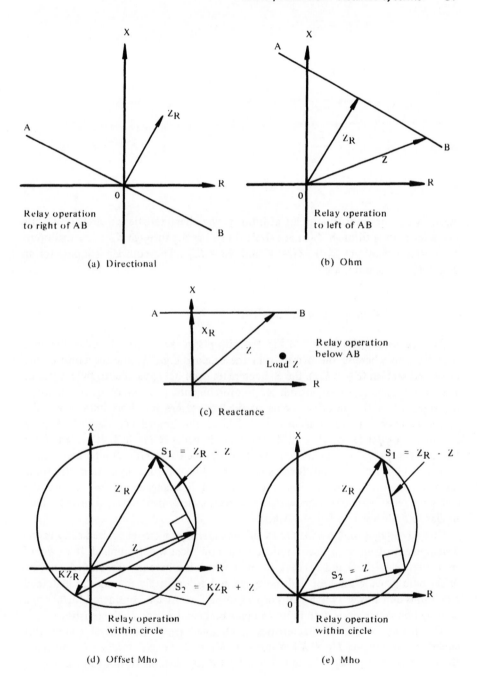

Fig. 9.4.4B   *Polar characteristics*

$$Z_{R1} = Z_{R2} = Z_R$$

giving

$$S_1 = IZ_R - V$$
$$S_2 = IZ_R + K_4 V$$

Thus, when

$$S_1 = 0, \ Z = Z_R.$$

When $S_2 = 0, \ Z = \dfrac{-Z_R}{K_4}$

which for $K_4 \longrightarrow 0, \ Z \longrightarrow \infty$

Again, with one measuring point at infinity, the characteristic is a circle of infinite diameter passing through Z_R, i.e. a straight line passing through Z_R (thus the inputs for an ohm relay are $S_1 = IZ_R = V$ and $S_2 = IZ_R$. The equivalent inputs for an amplitude comparator are

$$S_0 = 2IZ_R - V \ \text{ and } \ S_r = V$$

The *reactance relay* shown in Fig. 9.4.4B(c) is in fact a special case of the ohm characteristic whereby the offset from the origin is equal to the reactance of the protected section ($Z_R = X_R$), the characteristic itself being parallel to the resistance axis. Reactance relays were introduced to overcome the problem of arc resistance at the point of fault, the effect being particularly severe for short lines where the vector addition of arc resistance and faulted line impedance could cause the resultant impedance to fall outside the relay impedance reach. It is evident from Fig. 9.4.4B(c) that the reactance relay is susceptible to operation under load conditions and power swings, and for this reason must be used in conjunction with some form of monitoring relay. The most successful arrangement was to use the reactance relay with a voltage-restrained directional element having a characteristic similar to that shown in Fig. 9.4.4B(d).

The advantages of the *mho family of characteristics* were recognised early in the development of distance relays and they are now most widely applied. It is evident from Figs. 9.4.4.B(d) and (e) that the mho, which is inherently directional, and the offset mho provide the best fault coverage and are relatively immune to high load conditions and power swings. This aspect of performance is considered further in a later section which deals with more complex comparators and characteristics.

The family of mho characteristics is obtained from the general comparator equations by putting $K_1 = K_2 = K_4 = 1$, $K_3 = K$, i.e. $S_1 = IZ_R - V$ and $S_2 = KIZ_R + V$ giving $S_1 = Z_R - Z$ and $S_2 = KZ_R + Z$ (by dividing through by I).

9.4.4.1   Offset mho characteristic: The measuring points corresponding to

$S_1 = S_2 = 0$ are $Z = Z_R$ and $Z = -KZ_R$ where K determines the degree of offset impedance as a proportion of the forward impedance reach. Equivalent inputs to the amplitude comparator are

$$S_0 = IZ_R (1+K) \text{ and } S_r = IZ_R (K-1) + 2V$$

The *directional mho characteristic* is obtained by putting $K = 0$, i.e. $S_1 = IZ_R - V$ and $S_2 = V$ for the phase comparator, giving $S_1 = Z_R - Z$ and $S_2 = Z$ (by dividing through by I) and $S_0 = IZ_R$ and $S_r = 2V - IZ_R$ for the amplitude comparator.

9.4.4.2   Other characteristics: In practice, the range of characteristics described by the phase-comparator equations can be significantly extended by modifying the phase criterion for operation. Thus, a more general angular criteria would be:

$$\alpha_2 < \phi < \alpha_1$$

where ϕ is the phase angle between S_1 and S_2 and α_1 and α_2 are the fixed phase-angle limits for operation. The significance of this may be illustrated with reference to Fig. 9.4.4C(a) which shows the effect of phase-angle criterion on characteristic shape for the comparator inputs

$$S_1 = IZ_{R1} - V$$
$$S_2 = IZ_{R2} + V$$

The characteristics show clearly that the overall polar characteristic is made up of intersecting circular arcs. In the case of straight-line characteristics, the special case of the continuous (ohm) straight line in Fig. 9.4.4C(b) becomes a discontinuous characteristic made up of intersecting straight lines. This simple means of characteristic shaping has been used in practice particularly using a symmetrical phase criteria of less than $90°$ for a mho characteristic to produce the 'lens'-shaped characteristic of Fig. 9.4.4C(a).

It is, of course, possible to derive analytically the general equation for the polar characteristic of a phase comparator from the general comparator input equation and phase criteria but this is a tedious, although straightforward, process which is beyond the scope of this text. The equivalent analysis for an amplitude comparator is given in Chapter 6, Section 6.8.4.

9.4.5   Measuring characteristics of relay schemes

Various combinations of the characteristics described in Section 9.4.4. have for many years formed the basis of distance protection schemes. An example of how

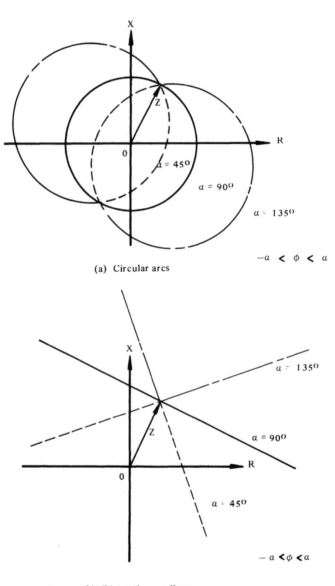

(a) Circular arcs

(b) Discontinuous lines

Fig. 9.4.4C   *Effect of varying phase criterion*

the main impedance and directional elements can be combined to provide three stages of measurement is illustrated by Fig. 9.4.5A. It will be noted that the Argand diagram has been superimposed on the feeder and orientated so that the latter can be represented as a line and as an impedance. For a stage-3 fault (F in Fig. 9.4.5A), the directional element operates, removes the control from the

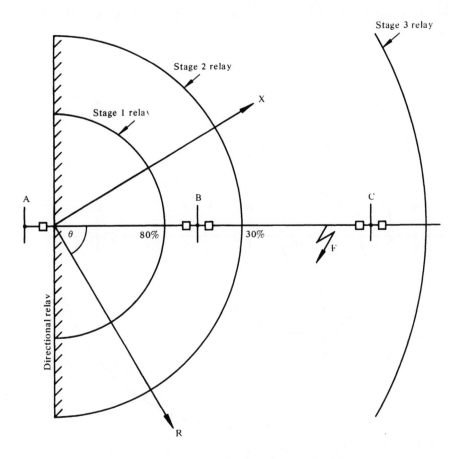

θ — Impedance angle of lines

Fig. 9.4.5A   *Relay characteristics for typical distance protection utilising plain impedance relays*

measuring relay and energises two timing relays (stage-2 and stage-3 time-lags). Because the fault is beyond the reach of stage-1 relay, no operation occurs. When the stage-2 time-lag runs out, the setting of the impedance element is increased to that of stage 2, but of course the fault is also beyond this reach. Only when stage-3 time-lag runs out and the setting adjusted to the stage-3 reach does relay operation occur and tripping of the breaker at A take place. This assumes that the fault is not cleared by the protection at B.

The scheme shown in Fig. 9.4.5B has been the most widely applied in the UK and in many other parts of the world. As shown, it is based on mho and offset mho relays and in one form a simple mho element is used for stages 1 and 2 and an offset mho element for stage 3. Tripping occurs immediately for faults within the stage-1 reach whereas for faults beyond stage 1, but within stage 2 or stage 3, tripping is controlled by timing relays which are energised by operation of the stage-3 relay.

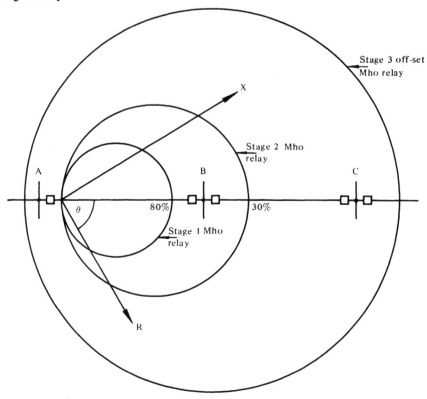

Fig. 9.4.5B   *Relay characteristics for typical distance protection utilising mho relays*

It is possible to have not only more stages of measurement but also separate measuring elements for each stage rather than switching the impedance reach of elements and indeed such arrangements are finding wide application because of advantages in carrier-aided schemes of distance protection (to be described).

The mho-based scheme continues to be widely applied but there is also increasing usage of more complex characteristics (Section 9.6.5). The polarised mho is a particularly important element of the mho-based scheme which was introduced as a means of overcoming certain performance limitations of the simple mho relay described earlier.

9.4.6   Mho characteristics

The directional mho relay, described by the input equation in Section 9.4.4, has limitations when required to operate for faults close to or at the relay location.

For such conditions, the voltage collapses to a small value which may be insufficient to give decisive relay operation. To overcome this for unbalanced faults it is sufficient to supplement the voltage applied as signal S_2 above, with a component which is derived from phases not directly involved in the fault. For balanced faults, all phases are obviously affected by the fault, and other special measures are necessary.

The use of additional healthy-phase voltage as described above results in a very important relay, known as the *polarised mho relay*, which is very widely applied and which warrants special attention. For many years, the effect of healthy-phase polarising voltage on the polar characteristic of the relay was not appreciated, it being assumed that the main effect was to ensure decisive operation at the relay location thus causing the characteristic to pass through the origin of the Argand diagram. The fact that such decisive operation was incompatible with marginal operation associated with the characteristic boundary was later realised in the 1960s and the true nature of the polarised mho characteristic was revealed through detailed analysis. Such analysis is beyond the scope of this text but because polarised relays hold such an important place in distance protection, a simplified means of deriving the polarised mho characteristic is described as follows:

The input signals necessary for a polarised mho relay, assuming a phase comparator is used, are,

$$S_1 = IZ_R - V$$
$$S_2 = KV + K_P V_P$$

where V_P is the voltage derived from a healthy phase or a combination of healthy and faulted phases and K and K_P are constants which can be complex. For example, an earth-fault relay connected for red-phase measurement may use a healthy blue-phase component suitably phase-shifted to be approximately in phase with the red-phase component. Alternatively, a component derived from the yellow-blue voltage may be used. A phase-fault relay connected for red-yellow measurement may use a component derived from the blue-red voltage, again suitably phase-shifted.

If now each of the equations above are divided by the current I, the following can be written, assuming $Z = V/I$:

$$S_1 = Z_R - Z$$

$$S_2 = KZ + \frac{K_P V_P}{I}$$

The component $K_P V_P/I$ may be expressed in terms of system source and line impedances (including zero-sequence components where phase quantities are involved) the particular expression depending on the healthy phase components used. Thus $K_P V_P/I = f(Z, Z_S)$ and the input equations can be rewritten:

$$S_1 = Z_R - Z$$

$$S_2 = K_1 Z + K_2 Z_S$$

Constants K_1 and K_2 are generally complex and for the purpose of further analysis it is convenient to include the argument $\underline{/\sigma}$ of K_1.

$$S_1 = Z_R - Z \qquad (7)$$

$$S_2 = K_1 \underline{/\sigma} Z + K_2 Z_S \qquad (8)$$

As was shown earlier, two points on the polar characteristic can be defined immediately by equating the expression for S_1 and S_2 to zero, thus defining marginal or boundary conditions as follows:

$$S_1 = 0, \text{ therefore } Z = Z_R$$

$$S_2 = 0, \text{ therefore } Z = \frac{-K_2}{K_1} \underline{/-\sigma} Z_S$$

If now the input equations are written as:

$$S_1 = Z_R - Z$$

$$\frac{S_2}{K_1} \underline{/-\sigma} = Z + \frac{K_2}{K_1} Z_S \underline{/-\sigma}$$

then it can be seen from Fig. 9.4.6A that any vector Z from the origin to D on to a line AB joining the extremities of vectors Z_R and

$$\frac{-K_2}{K_1} Z_S \underline{/-\sigma}$$

must represent the condition of cophasal quantities S_1 and

$$\frac{S_2}{K_1} \underline{/-\sigma}$$

because

$$AD = S_1 = Z_R - Z$$

and

$$BD = S_2 \frac{\underline{/-\sigma}}{K_1} = Z + \frac{K_2}{K_1} Z_S \underline{/-\sigma}$$

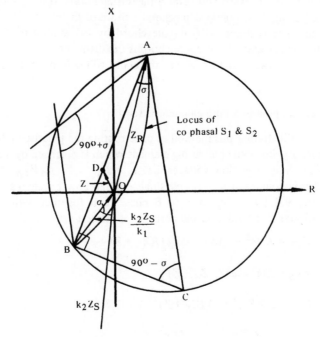

Fig. 9.4.6A *Geometrical construction of polarised mho (circular) characteristic*

Thus, if the boundary is defined by an angular criterion:

$$- 90° < \phi < 90°$$

where ϕ is the angle between S_1 and S_2, then, at the boundary, inputs S_1 and

$$\frac{S_2}{K_1} \underline{/-\sigma}$$

are displaced by

$$- 90° + \sigma, \text{ for } S_1 \text{ lagging } \frac{S_2}{K_1} \underline{/-\sigma}$$

and

$$+ 90° + \sigma, \text{ for } S_1 \text{ leading } \frac{S_1}{K_1} \underline{/-\sigma}$$

The diameter of the characteristic can be determined by drawing a line BC at right-angles to vector AB and a line displaced from AB by σ. The diameter is then AC since the chord subtends an angle $90° - \sigma$ at C.

In order to simply construct the characteristic the above can be summarised as follows:

(a)   Determine the forward and reverse measuring points from $S_1 = 0$ and $S_2 = 0$. A line joining these points is in general a chord of the circle.

(b)   The diameter is displaced from the chord by the argument of K_1 i.e. $\underline{/-\sigma}$. For the special case of $\sigma = 0$, the chord is the diameter.

In general, a line drawn at right-angles to the chord will determine the diameter.

9.4.7   Practical polarised mho characteristic

As a means of illustrating this method of constructing the polarised mho characteristic, consider a yellow-blue measuring element which is polarised by a combination of yellow-blue and red-yellow voltages, i.e. $S_2 = \overline{K}_1 V_{YB} + \overline{K}_2 V_{YR}$. It is assumed that the relay is connected to a simple single-end fed power system shown schematically in Fig. 9.4.7A feeding a Y to B phase fault together with the sequence impedance diagram. The voltages and currents of interest are:

$$V_{YB} = E[(a^2 - 1)Z - 1\cdot5Z_S]/(Z_S + Z)$$

$$V_{YB} = E(a^2 - a)Z/(Z_S + Z)$$

$$I_Y - I_B = E(a^2 - a)/(Z_S + Z).$$

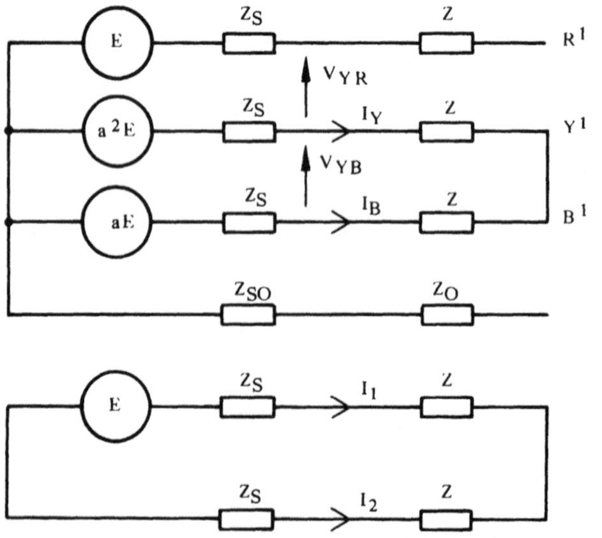

Fig. 9.4.7A   Equivalent circuits for single-end fed system with phase fault

In the equation for S_2, the constant K_2 includes a phase angle due to a phase-shift circuit and in this particular case an argument of $60°$ is chosen, i.e. $\overline{K_2} = K_2 \angle 60°$.

$$S_2 = E(a^2 - a)\, [K_1 Z + K_2 Z + \frac{\sqrt{3}}{2} K_2 \underline{/{-}30°}\ Z_S]\, /\, (Z_S + Z)$$

$$S_1 = E(a^2 - a)\, (Z_R - Z)\, /\, (Z_S + Z)$$

Each equation can be multiplied by $(Z_S + Z)\, /\, E(a^2 - a)$

$$S_1^1 = (Z_R - Z)$$

$$S_2^1 = (K_1 + K_2)\, Z + \frac{\sqrt{3}}{2} \underline{/{-}30°}\ K_2 Z_S$$

By equating S_1^1 and S_2^1 to zero to give two boundary points; we have:

$$Z = Z_R$$

and

$$Z = -\frac{\sqrt{3}}{2} \underline{/{-}30°}\ \frac{[K_2 Z_S]}{[K_1 + K_2]}$$

and because the argument of the coefficient of Z in S_2^1, i.e. the argument of $(K_1 + K_2)$, is zero, then the line joining the two vectors above is in fact the diameter of the characteristic.

Thus, in general, for all unbalanced faults in the forward direction the characteristic of the polarised mho relay is a circle enclosing the origin, the degree of reverse offset being determined mainly by the source impedance Z_S and the polarising constant K_2. For the special case of $Z_S = 0$, and also for balanced three-phase fault conditions, the characteristic always passes through the origin and when $Z_S = \infty$ the characteristic becomes a straight line.

The above appears to suggest a loss of directionality but in fact for reverse faults eqns. 7 and 8 become:

$$S_1 = -Z_R - Z$$

$$S_2 = K_1 \underline{/\alpha}\, Z + K_2 Z_S$$

because of the reversal of current, and so both boundary points defined by $S_1 = 0$ and $S_2 = 0$ are in the third or fourth quadrants and the characteristic is typically, as shown in Fig. 9.4.7B. The significant feature of the characteristic is that the origin is external to the boundary, and again only when $Z_S = 0$ does the boundary pass through the origin.

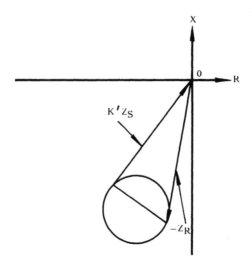

Fig. 9.4.7B   *Polarised mho characteristic-reverse power flow*

9.5   Development of comparators

Two-input comparators are essentially part of a more general class of multi-input comparators, but they have always held such an important position in distance protection that they warrant special consideration. The development of such comparators spans many years during which many different principles have been applied, the most successful of which are considered below.

9.5.1   Induction cup

As established earlier, there are basically two forms of distance measuring element known as amplitude and phase comparators, both of which may be categorised further according to whether electromechanical or electronic means are used to realise the device. Of the electromechanical types the most successful, which is still applied, is that based on the induction cup relay. Such a relay, which is described also in Chapter 6, is inherently a directional element in which the force relationship is given by $I_1 I_2 \sin \phi > K$. Thus the operating torque on the cup is proportional not only to the product of the applied currents but also to the phase angle between them and because I_1 and I_2 must always be positive the operation of the relay requires that the angle between the inputs be in the range 0-180°. Typical characteristics relating I_1, I_2 and ϕ are shown in Fig. 9.5.1A.

The induction cup element was certainly more efficient than the induction disc element, could work over a larger range of input quantities, and because of its much reduced inertia, was capable of greater operating speed. Also, unlike the two-input induction disc, there was very little interaction between the magnetic circuits.

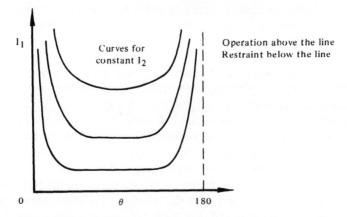

(a) Variation of current setting with angle

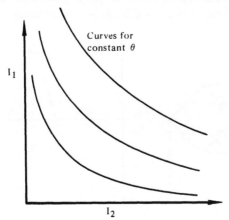

(b) Inverse relationships between input currents

Fig. 9.5.1A *Characteristics of induction cup relay*

Usually, in order to limit the torque produced at high inputs, a clutch mechanism was sometimes inserted between the contacts and the cup.

9.5.2 Rectifier bridge moving coil

Rectifier amplitude comparators represent the earliest application of semiconductors. The circulating current bridge arrangement shown in Fig. 9.5.2A(a) is still widely used and is therefore of special interest. As shown, the comparator comprises a polarised moving-coil relay and two bridge-rectifiers which are fed from the vector mixing circuits with the appropriate operate and restraint quantities. For cophasal inputs, with the operate signal very much less than the restraint signal, the voltage across the relay is limited to the small forward-biased diode level of the operate bridge-rectifier which passes the excess restraint current. As the operate current is increased, this excess current is decreased and reaches zero when $i_o = i_r$. Further increase in the operate level causes an increase in the relay current until the latter operates. Excessive operating current in the relay is avoided, this time by the shunting action of the bridge-rectifiers in the restraint circuit. The characteristic which best describes the above operation is illustrated in Fig. 9.5.2A(*b*) and shows the rapid transition from the restraint to operate condition. This transition is not as sharp for out-of-phase conditions but marginal operation is still defined by $i_o = i_r$ since this results in zero average current in the polarised element.

It is evident that the important feature of the bridge comparator described above is that the rectifier circuit which carries the smaller of the input currents acts as a nonlinear shunt across the relay and thus prevents excessively high values of current flowing through the relay. This allows a high range of working particularly with the recent developments in moving-coil elements which can now be extremely sensitive.

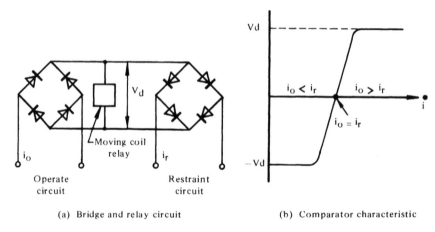

(a) Bridge and relay circuit   (b) Comparator characteristic

Fig. 9.5.2A   *Rectifier bridge amplitude comparator*

9.5.3   Electronic relays: introduction

Electronic relays have become established, particularly in the field of distance protection, because of significant benefits gained in terms of performance and application and also in terms of design and development. Electronic principles generally have afforded the manufacturer with a much greater development, design and manufacturing flexibility which is not apparent to the same degree with electromechanical equivalents. Such flexibility has resulted in a great many operating principles, only a few of which have been successful in practice having become firmly established in the light of exhaustive development and operating experiences. Now, when developments in protection and control are based on rapidly changing technologies such as components, computers, communications etc., it is inevitably difficult to maintain an up-to-date picture in any modern text on the subject. It is for this reason, that the following is concerned with successful and established principles which clearly illustrate certain fundamental aspects of design.

The earliest documented work on static elements, discounting Fitzgerald's investigations into carrier protection, was that of Wideroe who, in 1931, developed a number of relays including impedance-measuring types. These designs used thyratrons because of their high current-carrying capacity and triggering action and, although they did not find application, a number of fundamental operating principles were established for later designs.

In 1949, Macpherson, Warrington and McConnell developed a one-cycle mho distance relay using a pentode valve comparator and a thyratron for tripping.

In 1954, a new principle of measurement was described by Kennedy in which the main departure from previous investigations was the use of direct phase comparison and linear integration. This principle was later applied using semiconductor techniques and will be described in more detail later. Also in 1954, Bergseth described a relay based on the same principle of direct phase comparison but which used a telephone-type relay fed directly from the comparator. Although these later designs proved to be more stable (in the presence of extraneous pulses) than the earlier schemes based on instantaneous measurement, there was still some concern about the reliability of valves, the provision of power supplies and the cost and maintenance when compared with conventional relays. As a result of this, the application of electronic relays was limited only to field trials although a number of the principles established were later embodied in early transistor comparators.

The work on transistor comparators was started only a few years after the advent of the transistor. The potential of this device was soon recognised even though the types available initially were rather crude by present-day standards. There were inevitably questions concerning the reliability of these new components but over the years it became apparent that the developments in semiconductor technology and, indeed, component technology as a whole, were resulting in better and more precise designs. This progress has continued and at the present time there is increasing application of relays based on semiconductor electronics.

9.5.4 Comparator development

The most successful distance relays, developed in recent years using semiconductor technology, are based on phase rather than amplitude comparators. Two particular types are very widely applied, both in this country and overseas, and are known as the block-instantaneous and the block-average comparators. Both types illustrate important principles concerning comparator theory and are therefore considered in some detail in the following sections.

(*a*) *Block-instantaneous comparator:* The essential components of an impedance-measuring element based on the block instananeous principle are shown sche-

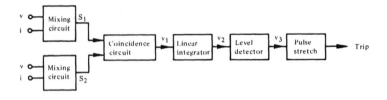

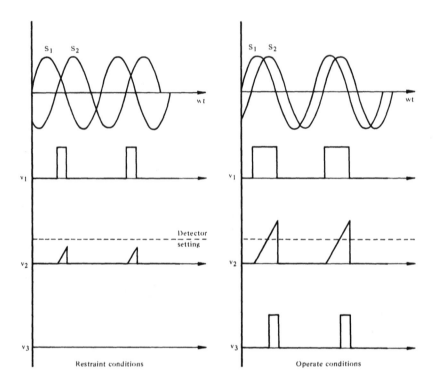

Fig. 9.5.4A   *Block instantaneous comparator and typical waveforms*

matically in Fig. 9.5.4A. In this scheme the mixing circuits are designed to provide voltages S_1 and S_2 appropriate to the desired characteristic as explained earlier. These voltages are applied to the coincidence circuit which provides an output V_1 only when the input signals are both, for example, of positive polarity, i.e. the output pulse-width is representative of the phase difference ϕ between S_1 and S_2.

This pulse is applied to the integrator, the output of which increases linearly to a level which is proportional to the input pulse-width. Thus, the angular criterion for operation is controlled by the time taken for the integrator output to reach the detector setting. For contemporary characteristics this critical angular displacement

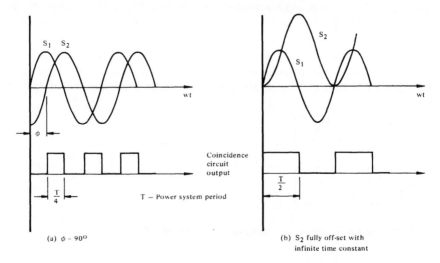

Fig. 9.5.4B *Comparison of steady-state and extreme dynamic conditions*

between S_1 and S_2 is arranged to be $90°$.

The above is a very simple description of the block-instantaneous principle based on steady-state operating conditions. In practice, the comparator system is required to respond very quickly (less than 1 cycle of power frequency) and at the same time ensure accurate impedance measurement even in the presence of high levels of transient. A particularly difficult measuring problem is caused by the presence of long duration, high-level d.c. transients mainly in the current waveform. The problem is best illustrated by comparing input waveforms under marginal conditions with and without d.c. offset components; the hypothetical (worst-case) of an infinite time constant 'transient' is considered, as shown in Fig. 9.5.4B. Although both conditions should ideally give rise to marginal conditions, it is evident that the block-instantaneous comparator will operate decisively for the transient case, this being known as transient overreach, i.e. under certain transient conditions the measuring element will respond to conditions *beyond* the operating boundary. This problem may be overcome by noting that if transient overreach occurs during, say, positive coincidence measurement, then underreach must occur during negative coincidence measurement; the converse applies. Thus by applying the constraint that both measurements should separately indicate a trip condition, transient overreach is avoided but considerable time-delay may be introduced because the transient component must decay to a level which allows both measuring channels to operate.

Contemporary schemes based on the block-instantaneous comparator overcome the dynamic measuring problems by filtering the input quantities, as shown typically by Fig. 9.5.4C. Thereafter, coincidence measurement occurs on both positive and negative half-cycles and the summated coincidence pulses applied to a simple integrating circuit. This additional input filtering has resulted largely from practical experience of early block-instantaneous comparators and tends to support the widely accepted belief that electronic comparators need some degree of (optimised) inertia equivalent to the mechanical inertia of the earlier electromechanical relays. It is true that comparators can be designed to operate almost instantaneously, allowing for delays due to point-on-wave of fault occurrence, but such designs, which effectively have no inertia, are generally unstable in the presence of the high

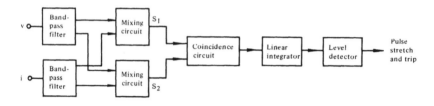

Fig. 9.5.4C   *Block instantaneous comparator with input filters*

levels of transients and interference typical of power systems. However, the required inertia must be controlled so as to allow high-speed, transient-free performance and it is the fact that this can be achieved which gives the electronic relay such an advantage over the electromechanical counterpart.

(*b*)   *Block-average comparator:* The block-average comparator is in many respects similar to the block-instantaneous type and indeed the block diagram of Fig. 9.5.4A applies, the essential difference being in the integrating function. This difference however results in a significant difference in the dynamic performance of the two systems with the block-average comparator providing a high-speed measuring system free from transient measuring errors, this being achieved by optimising the integrating (or inertia) function.

The block-average principle was first proposed by Wedepohl. Operation up to the output of the coincidence circuit is identical with that described in the previous Section. Combined measurement on both positive and negative half-cycles ensures good dynamic performance and the resulting comparator output pulses, which are representative of the phase difference between S_1 and S_2, are applied to the integrating circuit. The output of this circuit increases linearly at some prescribed rate when a pulse is applied to the input and falls, again at some prescribed rate, when the pulse ceases. The final element is a level detector which changes state when the integrator output exceeds some preset value and resets when the output falls below some second value. Typical waveforms for marginal (90° criterion) and operate conditions are shown in Fig. 9.5.4D, from which it is evident that there is an effective gain in integrator output only for the condition $\phi < 90°$. The rise and fall rates define the angular criterion for operation and may be set to any desired value. Cyclic switching or relay 'chatter' for marginal phase displacements is avoided by ensuring that the setting and resetting values for the level detector are such as to accommodate the peak-peak value of the integrator ripple.

To ensure good transient performance of the block-average comparator, the

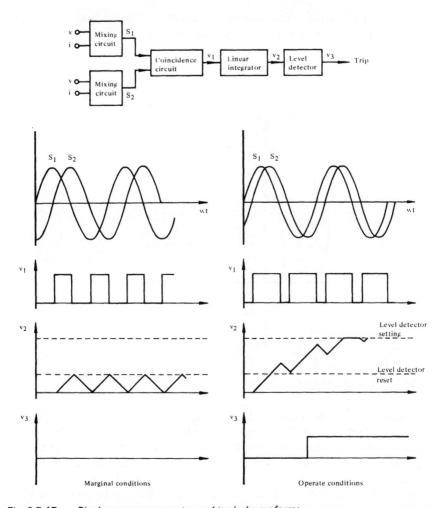

Fig. 9.5.4D   *Block average comparator and typical waveforms*

initial rate-of-rise of integrator output is limited to some minimum value, defined by the conditions examined earlier with reference to Fig. 9.5.4B. Thus, assuming a 90° operate criterion, the steady-state marginal condition is characterised by a pulse train of unit mark/space ratio, the pulse-width being one-quarter of the power system period. The hypothetical maximum transient condition shown in (*b*) also corresponds to unit mark/space ratio but the pulse-width in this case is one-half of the power system period. In order that both pulse-trains should correspond to marginal conditions it is necessary to ensure that the rate-of-rise of integrator output is less than $2V_d/T$ where V_d is the detector setting and T is the power system period. This constraint allows a minimum operating time of $T/2$ for co-phasal inputs but also ensures that there is no gain in integrator output for the conditions considered in Fig. 9.5.4B.

Stability in the presence of extraneous pulses is assured by the integrating action described above. Indeed, it is evident that the integrator and level detector allow considerable control over the dynamic performance. Delay due to point-on-wave of fault occurrence varies from 0 for cophasal inputs to $T/2$ at the boundary of operation.

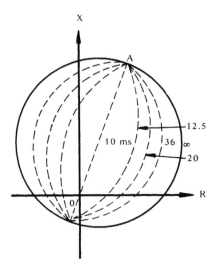

Fig. 9.5.4E   *Theoretical polar timing contours*

It is apparent from the above that in the case of the block-average comparator there is a distinct relationship between operating-time and phase difference of comparator input quantities, there being in addition some variance due to point-on-wave effects. This relationship can be used to show operating times relative to the measuring characteristic. The way in which this is done will be described later in relation to presentation of performance but the principle can be illustrated with reference to Fig. 9.5.4E. The polar boundary corresponds to a phase relationship between relay inputs of ± 90°. Thus constant-angle contours, corresponding to angles less than ± 90°, are arcs of circles with the characteristic impedance Z_R as a chord as shown. These arcs can be identified with constant-time contours which are derived from the theoretical time-angle relationship of the measuring element, the derivation of which is beyond the scope of this work.

9.5.5   Practical realisation of static phase comparators

The essential components of the block diagram shown in Fig. 9.5.4A are considered below in terms of circuit diagrams.

(a)   Derivation of relaying signals

Contemporary relay characteristics usually require complex addition and/or subtraction of voltage and current signals at the comparator input terminals. This may be achieved by the use of current transformers with multiple primary windings and a single secondary winding. The design of such transformers is complicated by many factors but the main disadvantage is the number of iron-cored elements required in a full distance protection.

Signal mixing required for phase comparators can be more conveniently achieved by means of the summation amplifier, the design of which is now well established, having been well tried in such fields as analogue computing and control systems and for which linear integrated circuit (i.c.) amplifiers greatly simplify the realisation. Fig. 9.5.5A shows a practical arrangement in which the linear amplifier is characterised by high open-loop gain and a virtual-earth input so that the overall gain is defined by the input resistors and the feedback resistor. Input signals V_1 and V_2 are typically derived from current and voltage transformers.

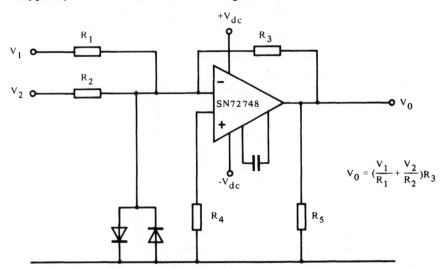

$$V_0 = (\frac{V_1}{R_1} + \frac{V_2}{R_2})R_3$$

Fig. 9.5.5A   *Summation of a.c. signals*

Because of the particular application to phase comparators, any amplitude distortion of the amplifier output can be arranged to be of little significance and in fact the mode of operation is such that the amplifier is overdriven except in the region of marginal relay operation. It is immediately apparent, however, that phase distortion should not be allowed to occur.

The impedance converting properties of the amplifier (high-input impedance – low-output impedance) are advantageous and the mixing of the input signals can be accurately defined using close-tolerance resistors R_1 and R_2. Additional phase-shifting requirements can be accommodated by introducing capacitance or inductance into the input circuit.

(b)   Phase-detecting circuits

The essential phase-detecting function has been realised in many different ways in practical comparator systems.

Coincidence circuits: The phase comparison of two or more signals may be effected by determining the period of their instantaneous polarity coincidence. In practical terms this is achieved by means of an 'AND'-gate circuit, a typical example

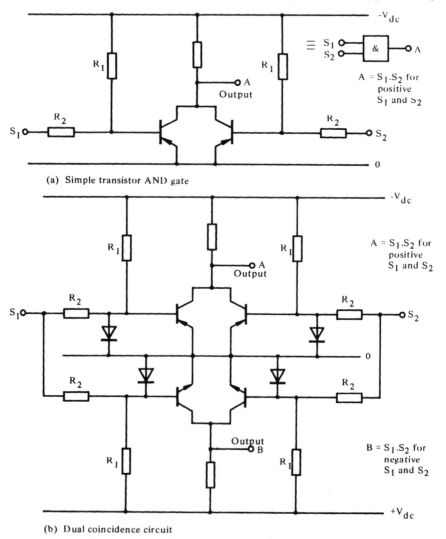

(a) Simple transistor AND gate

(b) Dual coincidence circuit

Fig. 9.5.5B   *Phase comparison based on polarity coincidence*

of which is given in Fig. 9.5.5B(a). With the symmetrical arrangement shown, both transistors are conducting, in the absence of input signals, by virtue of the base current drive through R_1. It is therefore necessary for *both* inputs to be simultaneously positive and to exceed the current in R_1 in order for the common-collector output voltage to change. Coincidence detection on both positive and negative half-cycles requires an arrangement such as shown in Fig. 9.5.5B(b) in which negative polarity coincidence is detected by the parallel combination of *n-p-n* transistors. The outputs from points A and B are subsequently summated.

With modern circuit techniques the setting of the AND-gate, as determined by the circuit parameters, can be made extremely small but, in practice a finite well defined setting is usually preferred to ensure a definite restraint condition when S_1 and/or S_2 is equal to zero.

The effects of this setting can be illustrated by noting that for impedance-measuring relays, one of the inputs, S_1, comprises an operate component, IZ_R, and a restraint component, V, such that $S_1 = IZ_R - V$. If V_S is the relay setting then $IZ_R - V = V_S$ corresponds to a point* on the operating boundary and by normalising this expression to IZ_R the following is obtained:

$$\frac{V}{IZ_R} = 1 - \frac{V_S}{IZ_R}$$

The term V/IZ_R is shown later to be a measure of relay-measuring accuracy which should ideally be maintained at unity over a wide range of signal levels, but obviously at the lower signal levels the error term V_S/IZ_R can be appreciable. This effect is shown in Fig. 9.5.5C, in which the limits of assigned accuracy are indicated by lines a and b so that the useful working range of the relay is limited. The ideal characteristic in Fig. 9.5.5C can be approached by compensating for the finite relay sensitivity, i.e. by preventing the restraint signal from being effective until the comparator setting is reached. One convenient method of compensation is shown in Fig. 9.5.5D in which a nonlinear impedance comprising semiconductor diodes is connected in the restraint circuit of the summation amplifier input. The amplifier feedback resistor can be adjusted to provide close matching of the well defined diode voltage to the coincidence circuit setting.

Another form of phase-detecting circuit is the rectifier bridge type which is described fully in Chapter 6, Section 6.8.7.

(c)   Output integrator and level detector

Block comparators allow a freedom of design by virtue of the variable pulse width output which is a measure of the phase difference between input quantities under

---

*When IZ_R and V are cophasal, it can be shown that the r.m.s. components of the input S_1 can be equated to the d.c. setting V_S, assuming a relay operating criterion of $\pm 90°$ rad.

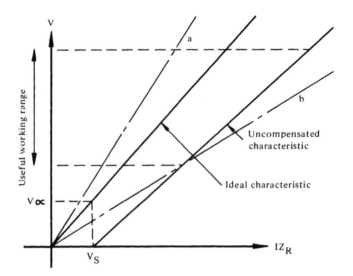

Fig. 9.5.5C   *Effect of finite comparators setting on relay range and accuracy*

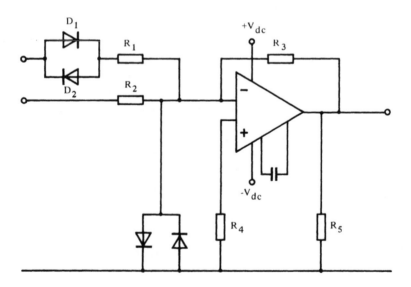

Fig. 9.5.5D   *Method of compensating for finite sensitivity*

steady-state conditions. Under extreme transient conditions, the nature of the output pulse-train is changed and this allows design criteria for the output circuit to be established which will ensure freedom from transient effects. A typical output stage uses a pulse-width to amplitude converter and level detector. The converter is in fact a linear integrator, an example of which is shown in Fig. 9.5.5E in the form

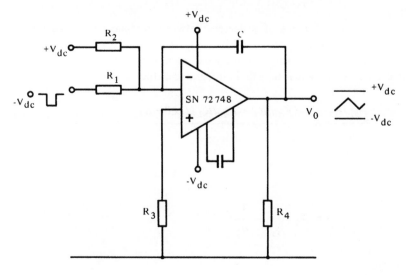

Fig. 9.5.5E   *Linear integrator*

of a high gain amplifier to which is applied capacitive feedback via C. It is arranged that

$$R_1 < R_2$$

This ensures that the amplifier is fully conducting so that the output voltage and hence the charge on C is virtually zero. When a pulse appears at the integrator input, corresponding to the coincidence period between comparator inputs, the integrator output increases linearly at a rate given by

$$V_0 = \frac{V_{dc}/R_1 - V_{dc}/R_2}{C}$$

On cessation of the pulse, corresponding to the noncoincidence period the integrator output falls linearly at a rate

$$\frac{V_{dc}/R_2}{C}$$

It is evident that the ratio of 'rate-of-rise' to 'rate-of-fall' of the integrator output defines some critical pulse-width, and hence phase angle, at which there is a gain in output which will operate the level detector. Thus, the angular criterion for relay operation is controlled by R_1 and R_2, the quantitative relationship being derived as follows:

For the phase difference of ϕ degrees between comparator inputs, the pulse-width applied to the integrator is $(180 - \phi)$ degrees, and for an angular criterion α degrees, the mark/space ratio for marginal operation is $(180 - \alpha)/\alpha$. Also, under marginal conditions, there is no gain in integrator output so that

$$\frac{V_{dc}/R_2}{C} \alpha = \frac{(V_{dc}/R_1 - V_{dc}/R_2)(180 - \alpha)}{C}$$

or

$$\frac{\alpha}{180 - \alpha} = R_2 \left(\frac{1}{R_1} - \frac{1}{R_2}\right)$$

from which

$$\frac{R_2}{R_2} = \frac{180}{180 - \alpha}$$

Actual values for R_1, R_2 and C are determined by the required operating time. It is evident from the above that the simple circuit of Fig. 9.5.5E has a profound effect on both the steady-state and dynamic performance of the overall scheme.

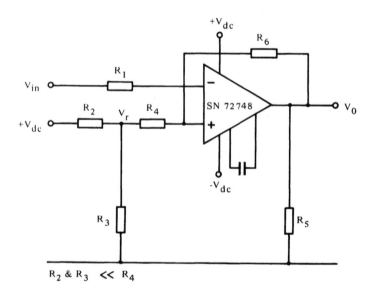

Fig. 9.5.5F   *Level detector*

Details of a *level detector* based on the well known Schmitt Trigger circuit are given in Fig. 9.5.5F. As stated, special consideration must be given to the setting and resetting levels of the detector insofar as they must be related to the integrator output ripple. For a relay having a 90° phase-angle criterion, the detector setting must exceed the level reached by the integrator output in a quarter of the periodic time. The difference between the set and reset level must also exceed this level if relay 'chatter' is to be avoided in the region of the operate boundary. Thus, a practical design has a setting of two thirds of the integrator excursion limit and a reset level of one third, which, for an integrator response time of not less than one half of the system period, is considered to be an optimum design. The reset time for such an arrangement is equal to the basic integrator response time.

9.6   More complex relaying characteristics

9.6.1   Basis for shaped polar characteristics

For many years, multizone impedance-measuring elements having circular characteristics (both polarised and offset mho) have been favoured for the protection of high-voltage networks largely because of their simplicity in terms of physical realisation and application. The use of the polarised mho element is even more firmly established now that there is a better understanding of the effects of sound-phase polarising voltages. It is now possible, however, to realise more complex characteristics, which are more suited to particular applications, by virtue of the flexibility afforded with electronic circuit developments, such characteristics being uneconomic using traditional relaying techniques. The provision of such characteristics must be tempered with consideration of the system protection requirements and the complexity, cost and performance of the relay which leads usually to a best overall compromise having regard for any conflicting requirements.

Generally, the requirements for the various zones of protection can be summarised as follows:

(a)   High-speed clearance of faults in zone 1 of the protected circuit, preferably from all terminations, and time-graded clearance for other zones.

(b)   High stability against tripping for all faults outside the relaying polar boundary and particularly for faults behind the relay location where directional measurement is concerned; a back-up zone of protection may be included which has a well-defined reverse reach.

(c)   High-stability against through-load and power-swing conditions and against healthy-phase impedances whilst maintaining coverage for resistive faults.

(d)   A controlled reach in the fourth quadrant of the Argand diagram so as to accommodate close-in arcing faults which are fed also from remote sources.

In the case of multi-element schemes, considerations such as the above may lead to different characteristics being used for phase and earth faults, e.g. circular characteristics for phase faults and reactance or trapezoidal characteristics for earth faults.

The latter represents a significant departure from the traditional circles and straight lines and is one example of a special characteristic which is finding wide application in contemporary schemes of distance protection. There are a number of ways of achieving discontinuous characteristics and the most common are considered below in relation to phase-angle comparators.

9.6.2   Change of angular criterion

The familiar straight-line and circular characteristics are defined by an angular operating range of 180°, and usually the phase-angle criterion is symmetrical, i.e. $-90° < \phi < 90°$. Characteristics comprising arcs of circles or intersecting straight lines are obtained when this angular criterion is changed. Typical examples of such characteristics are given in Fig. 9.4.4C.

In practice, a change of operate criterion is rarely sufficient to produce the required characteristic and is usually applied in conjunction with the multicomparator or multi-input techniques described below.

9.6.3   Multicomparator schemes

Multicomparator schemes have been used for many years to provide load blocking and directional reactance-measuring arrangements. They can, of course, be used to produce many complex relaying boundaries but such application must be tempered with careful consideration to the complexity or economics of the final arrangement.

The basic principle is straightforward and may be illustrated by considering a reactance relay which is provided with a directional feature as shown in Fig. 9.6.3A. The final arrangement comprises the reactance-measuring element and, typically, a mho element with the output contacts connected in series, i.e. an 'AND'-gate configuration. The final characteristic is therefore the area common to the constitutuent characteristics as shown.

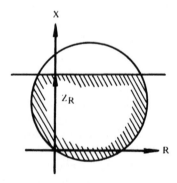

Fig. 9.6.3A   *Example of directional reactance characteristic (multi-relay)*

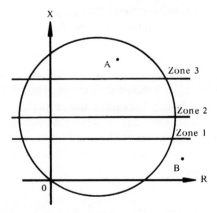

Fig. 9.6.3B   *Three-zone directional reactance scheme*

For certain applications, the principle requires careful assessment with regard to time co-ordination of output contacts. The scheme shown in Fig. 9.6.3B illustrates the difficulty where the main protection comprises a three-stage reactance arrangement and a mho starting relay. It is assumed that a fault at point A, just beyond the reach of zone 3, is cleared elsewhere in the system and that conditions revert to those described by vector OB. In this case it is possible for the zone 1 reactance element to operate before the mho relay resets and it is therefore possible for maloperation to occur.

Time co-ordination problems often arise where the overall boundary, either wholly or partially, comprises straight-line characteristics which are 'AND'-gated because these inherently have operating zones which extend beyond the final characteristic boundary. An alternative approach, which does not have this drawback, is illustrated in Fig. 9.6.3C. The final characteristic is seen to comprise intersecting circular characteristics which have well defined operating zones, the

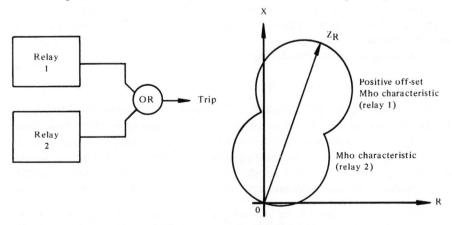

Fig. 9.6.3C   *Multi-relay arrangement and resultant characeristic*

output of the separate relay elements being 'OR'-gated.

More recently the multicomparator principle, based on solid-state phase comparator elements with modified phase criteria, has been used to realise the quadrilateral characteristic as illustrated by Fig. 9.6.5A(f).

This approach, because of the design freedom associated with solid-state tech-·niques, has tended to be more successful than the equivalent schemes based on electromechanical elements largely because of size reduction and the fact that more complex measuring characteristics can be achieved with high performance and acceptable cost. In many cases however the multi-input comparator (both electromechanical and solid-state types) has afforded an elegant solution in achieving more complex measuring boundaries.

9.6.4   Multi-input comparators

Early designs of multi-input comparators include six and eight-pole induction cup elements and multi-input versions of the bridge rectifier comparator, the latter being capable of producing characteristics comprising conic sections. More recent investigations have concentrated on multi-input phase comparators for characteristic shaping in which all inputs are 'AND'-gated; this may be effected typically as shown in Fig. 9.6.4A. The effect of combining inputs in this way may vary according to the type of comparator used, i.e. whether it is of the pulse* or the block type. This is not true for two-input types where, if the inputs are identical, then the steady-state characteristics are also identical. For more than two inputs, however, it must be remembered that for the block comparator, the width of the output pulse is defined by the signals which are most remote in phase and that any pair of signals may fulfil this condition. Thus, all combinations of signal pairs and their characteristics must be considered and the resultant characteristic taken as the common area. In general, a $90°$ comparator having n inputs will produce a resultant characteristic which is the area common to $n(n-1)/2$ separate two-input characteristics. Not all of the characteristics necessarily contribute to the resultant, i.e. some may be redundant.

As an example of this, consider the three inputs

$$S_1 = IZ_R - V, S_2 = IZ_R, S_3 = IZ_R + V$$

applied to the block-type comparator. The resultant characteristic is found by taking the area common to the characteristic of each pair of inputs in turn. Thus inputs S_1 and S_2 produce an ohm characteristic with the characteristic impedance in the first quadrant; inputs S_1 and S_3 produce an impedance characteristic; inputs S_2 and S_3 produce an ohm characteristic with the characteristic impedance in the third quadrant. In this particular case the ohm characteristics do not limit the

*The pulse-type of comparator is one in which a narrow pulse is derived at some instant in the cycle of one input, say S_1, and the polarity of the second input, say S_2, is determined at the instant the pulse occurs.

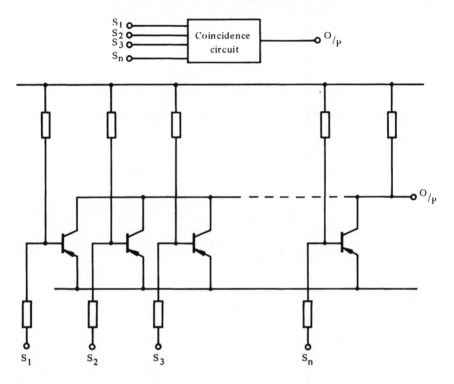

Fig. 9.6.4A   *Multi-input AND gate*

coverage of the impedance characteristic.

The flexibility of the multi-input comparator in terms of the range of characteristics available can be further enhanced by appropriate phase-shifting of current and voltage signals and in some cases by modifying the phase-angle criterion for operation. An example of how signal phase-shifts can be used to modify a characteristic is shown in Fig. 9.6.4B. Thus (*a*) illustrates the characteristic which results from the three inputs:

$$S_1 = IZ_R - kV$$
$$S_2 = IZ_R$$
$$S_3 = V$$

It is assumed that the angle of Z_R (the characteristic angle) is $20°$. With this input arrangement, the mho characteristic is seen to be unaffected by the ohm (due to S_1 and S_2) and the directional (due to S_2 and S_3) characteristics.

If now the input signals are modified to be

$$S_1 = IZ_{R1} - kV \angle{-10°}$$
$$S_2 = IZ_{R2}$$
$$S_3 = V$$

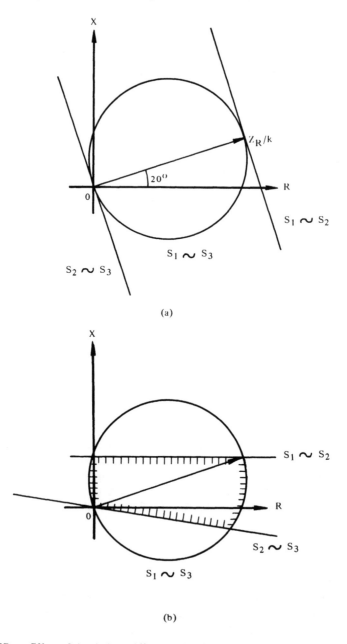

(a)

(b)

Fig. 9.6.4B   *Effect of signal phase shifts on polar characteristic*

Where the angle of Z_{R1} is $10°$ and that of Z_{R2} is $80°$ then the resultant characteristic is as shown in (*b*).

9.6.5 Alternative characteristics

Examples of alternative directional impedance measuring characteristics which have been applied in recent years are shown in Fig. 9.6.5A. All of these characteristics can of course be provided with a reverse offset if required.

The characteristics shown in (*a*), (*b*) and (*c*) are particularly useful where discrimination against power swings or healthy-phase encroachments are desirable. The elliptical characteristic (*b*) which has not been widely applied, may be derived from a 3-input amplitude comparator whereas the characteristic of (*c*) requires a 3-input phase comparator to produce a boundary representing a continuously variable phase criterion related in a controlled manner to the primary system quantities of voltage, current and phase angle. Trapezoidal (or quadrilateral) characteristics, such as shown in (*f*) may be derived using either 2 or more comparators or a single 3-input comparator, usually with a modified angular criterion for operation. The multicomparator arrangement may sometimes be used to give the added flexibility of independent X and R adjustments.

Where an extended reach along the resistive axis is required then the characteristics of (*d*), (*e*) and (*f*) may be applied. The combination of the plain impedance characteristic with polarised mho starting shown in (*d*) has effectively the coverage of an impedance relay particularly when the effects of crosspolarisation on the mho characteristic are taken into account. A disadvantage exists with such an arrangement which was mentioned in Section 9.6.3, i.e. the use of two relays to produce a composite characteristic creates problems of co-ordination of response times of the two relays.

An alternative to that shown in (*d*) is the characteristic of (*e*) which is in fact the combination of reactance and directional mho mentioned in Section 9.6.3. Here again the problem of time-coordination exists although modern circuit techniques have been used to combine comparators in a manner which yields the characteristic of (*e*) and which also emphasises the effects of crosspolarisation of the directional element so that significant extension of the resistive reach is obtained under unbalanced fault conditions. From this standpoint, such a characteristic has an advantage over the quadrilateral type insofar as the latter is a compromise between maximum resistive coverage and the avoidance of load and power swing encroachment. Full benefit is gained from quadrilateral characteristics only if independent control of resistance and reactance components of characteristic impedance can be achieved. This is not usually possible with multi-input comparators and requires multicomparator configurations.

9.7 Presentation of performance

9.7.1 Requirements

To define precisely the dynamic performance of a distance relay so that its

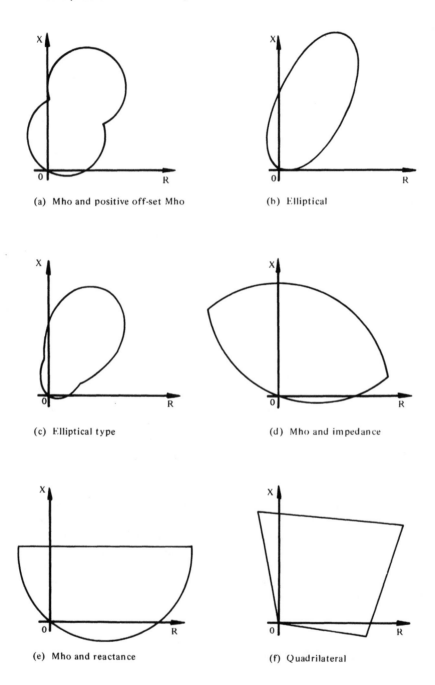

(a) Mho and positive off-set Mho

(b) Elliptical

(c) Elliptical type

(d) Mho and impedance

(e) Mho and reactance

(f) Quadrilateral

Fig 9.6.5A   *Special polar characteristics*

application to a practical power system is meaningful is somewhat complicated by the number of performance parameters of interest and the variability of the energising quantities. This is not so true of the steady-state performance which is relatively easily defined in terms of measuring accuracy over the full working range of steady-state current, voltage, and phase angle. However, a complete description of performance must take account of the following:

(a) the accuracy of impedance measurement and how this accuracy varies with the fault level (MVA)

(b) how the accuracy varies with the degree of d.c. and a.c. transient components in the fault current and voltage

(c) how the speed of operation varies with the distance to fault and the fault level

(d) how the speed of operation is affected by transient components in the current and voltage

(e) how the performance, particularly the speed of operation, is affected by extraneous fault resistance

(f) the stability of the relay for faults behind the relay location and on other phases.

There are additional factors which must be established and declared to the user concerning the influence of surges and interference and also temperature, frequency etc., on the basic performance parameters defined above.

9.7.2 Display of measuring accuracy

It is evident that the display of information is difficult because of the number of variables involved. However, the basic parameters which define the performance of a distance relay are the speed of operation and the input voltage and current. The current and voltage can be related back to the primary system via the current and voltage transformers and from the simple single-phase representation in Fig. 9.7.2A.

$$I = \frac{E}{Z + Z_S}$$

and

$$V = E \frac{Z}{Z + Z_S}$$

For a practical power system, conditions are such that at any given time the source MVA (and hence Z_S) and the length of protected line are known, the only variable being the fault position. The source MVA may of course vary with time depending on the amount of generating plant behind the relay location. Thus, information is required as to the impedance of the line beyond which the relay ceases to operate (relay cut-off point) and how this cut-off varies with source impedance Z_s. It is preferable to measure the variation of line impedance Z against the fixed reference

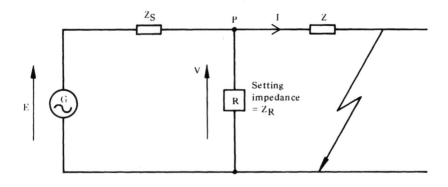

P	-	Relaying point	E	-	Normal system voltage
Z	-	Line impedance	I	-	Current at relaying point
		up to fault	V	-	Voltage at relaying point
Z_S	-	Source impedance	R	-	Measuring relay

Fig. 9.7.2A   *Single-phase representation of power system*

of the relay impedance setting Z_R. The ratio of measured cut-off impedance to relay setting impedance is defined as the relay accuracy x, i.e. $x = Z/Z_R$. Also the ratio of the source impedance to the relay setting impedance is called the System Impedance Ratio (S.I.R.) y where $y = Z_S/Z_R$.

The above expressions for V and I become:

$$V = E \frac{x}{x+y} \tag{1}$$

and $$I = \frac{E/Z_R}{x+y} \tag{2}$$

Expressions for x and y are:

$$x = \frac{V}{IZ_R} \tag{3}$$

and $$y = \frac{E-V}{IZ_R} \tag{4}$$

These two variables represent the per-unit fault position at the relay cut-off (x) and the system impedance ratio (S.I.R. or y) and are used to present basic performance data by plotting x directly against y. Ideally, x should be unity over a practical working range of source impedance corresponding to S.I.R.s up to 30 or more. This represents a particularly difficult condition which may be illustrated by considering the measuring input of a practical phase comparator to be of the form

$S_1 = IZ_R - V$ as defined earlier in Section 9.4.3.

Assuming it is necessary to determine the measuring accuracy at the characteristic angle (assumed equal to the line angle) and that the relay has a finite setting of V_S rather than infinite sensitivity, then for marginal operation

$$IZ_R - V = V_S$$

i.e.
$$1 - \frac{V}{IZ_R} = \frac{V_s}{IZ_R}$$

or
$$1 - x = \frac{V_S}{E} (x+y)$$

Substituting from eqns. 3 and 2. Therefore,

$$x = \frac{1}{1+V_S/E} (1 - \frac{V_S}{E} y)$$

Assuming $V_S \ll E$, then $x = 1 - \dfrac{V_S}{E} y$.

The form of this characteristic is shown by Fig. 9.7.2B from which it is apparent that the effect of a finite sensitivity is to cause a reduction in measuring accuracy.

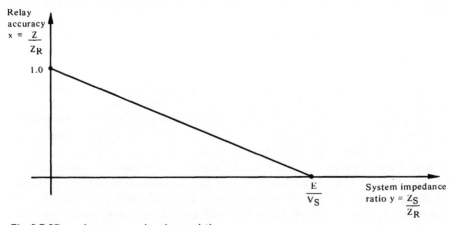

Fig. 9.7.2B   *Accuracy – s.i.r. characeristic*

If, for example, V_S is 1% of the nominal voltage E then the cut-off S.I.R. is 100 and the fault-position at cut-off for an S.I.R. of, say, 10 is 0·9. It is, of course, possible to increase the relay sensitivity so that the change in cut-off with S.I.R. is negligible over a practical working range. Alternatively, a finite sensitivity may be retained to ensure measuring stability and measures taken to compensate for its effect as

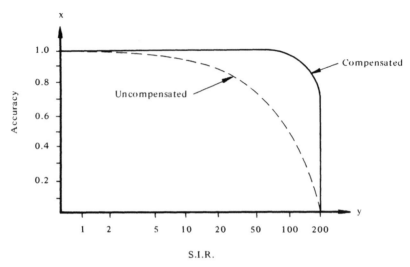

Fig. 9.7.2C   *Compensated characteristic*

described briefly in Section 9.5.5.

The presentation of x plotted against y is used to describe the accuracy capability of distance relays, although in practice it is convenient to use x against log y which takes the form typically as shown by Fig. 9.7.2C (shown for a compensated relay). Also, such a presentation would assume a particular point-on-wave of fault initiation (constant transient level), a constant value of fault resistance, if any, and so on.

9.7.3   Display of operating time

The variation of cut-off impedance (or accuracy) with system conditions as defined above does not fully describe the dynamic performance of a distance relay. It is necessary to know also the operating-time of the relay as a function of both fault-position and system-source conditions although, very often, this speed is assumed constant when describing such relays in simple terms. For example, the time-distance curves described in Section 9.3.1 assumed a constant low operating-time for zone-1 relays which extended to 80% of the protected line. Similarly, a longer constant time was assumed for zone-2 relays up to 120-150% of the first feeder, and so on. In practice, the operating time will depend very much on the operating principle of the measuring comparator and on the way in which input signals to it are derived. For example, there will be delays due to point-on-wave of fault initiation and variations due possibly to transient components in the input signals and, in the case of averaging comparators, a controlled variation in speed as the boundary of operation is approached. This important information can be combined with the accuracy presentation of Fig. 9.7.2C by noting that the boundary of operation in fact corresponds to a curve of infinite operating time, i.e. operation of the relay is marginal. All points within this boundary must correspond to finite operating time and by determining a number of such points a series of constant-time curves can be plotted as shown in Fig. 9.7.3A. Thus, as stated, the outer curve represents the boundary between operation and non-operation while inner curves give decreasing

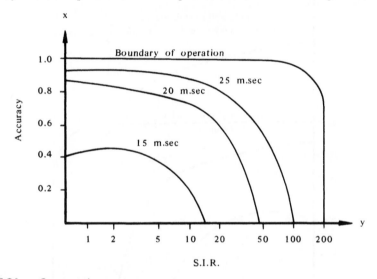

Fig. 9.7.3A   *Constant time contours*

operating times as the origin is approached and inputs to the relays increase. The time of the operation for a particular set of system conditions is obtained directly from the graphs by finding the fault position and the S.I.R. corresponding to the available source MVA and interpolating between contours.

The contours described above can be extended to accommodate reset-impedance boundary curves and reset-time contours.

9.7.4   Application of contour timing curves

The contour method of describing dynamic performance can be extended to cover a complete distance protection comprising a number of relays with different impedance settings, corresponding to particular zones, some of which are associated with additional time-lag elements. When applied in this way, the nominal impedance used in determining range and accuracy is taken as that corresponding to the complete length of the protected circuit. All relay characteristics are then plotted on this basis. Overall timing contours are assessed from the individual contours for each relay from which composite curves can be drawn as shown in Fig. 9.7.4A.

Since the performance of the overall protection may be different for different types of fault it will normally be necessary to have a series of diagrams covering the

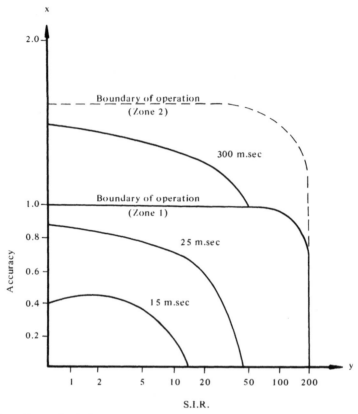

Fig. 9.7.4A   *Composite timing contours*

principal types of fault condition such as phase-earth, phase-phase, and three-phase. Further curves may also be needed to cover various transient conditions. Normally, two sets of curves corresponding to the maximum and minimum transient suffice.

The three-phase fault condition is of particular interest because of a problem associated with many forms of distance protection whereby the directional feature fails for faults at or near the relaying point. This problem was referred to in Section 9.4.5 in relation to directional mho characteristics where it was shown that a collapse of the input signal S_2 (the polarising voltage quantity) to zero rendered the comparator inoperative. The effect of this in terms of the contour curve presentation can be seen directly because additional contours representing constant relaying-point voltage can be superimposed on the accuracy – S.I.R. contours. Thus, because relaying-point voltage is

$$V = E \frac{Z}{Z_S + Z}$$

then $\qquad \dfrac{V}{E} = \dfrac{x}{x+y}$

from eqn. 1 of Section 9.7.2, i.e. a contour representing a constant per-unit value of voltage is determined by assuming values of the fault position x and determining corresponding values for y. The resultant diagram is illustrated by Fig. 9.7.4B.

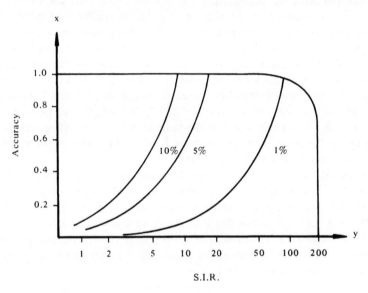

Fig. 9.7.4B *Constant voltage contours superimposed on accuracy — s.i.r. contours*

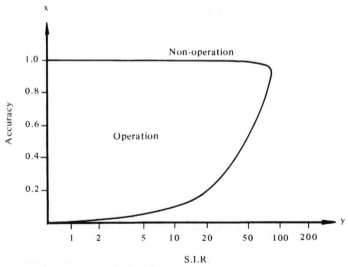

Fig. 9.7.4C *Mho relay; three-phase fault coverage*

Thus, if the voltage sensitivity of the input to which S_2 is applied is known, this can be related to a corresponding system voltage contour to give the resultant relay cut-off boundary shown in Fig. 9.7.4C. The contour shows the fraction of line which is unprotected for particular source conditions on the power system, such information being particularly difficult to obtain from other methods of performance presentation.

The contour form of presentation has the advantage that the constant time contours are independent of the relay setting. This is readily seen by considering any point on a contour curve which is defined by the parameters x, y and time t. From eqns. 1 and 2 of Section 9.7.2, the relay current corresponding to this point is $I = E/Z_R(x+y)$ and the voltage is $V = E\,x/x+y$. If now a current transformer of ratio $1:k$ is introduced then the new relay setting is Z_R/k, and if x and y remain unchanged the new source impedance becomes yZ_R/k and the new line impedance xZ_R/k. The fault current now becomes $I = E/(xZ_R/k + yZ_R/k)$ and the relay current becomes I/k $(= E/Z_R(x+y))$ which is the same as before the setting was changed. The voltage at the relay point is determined by x and y and, as far as the relay performance is concerned, there is no change.

If the setting of the relay is changed by inserting a voltage transformer or by modifying a VT ratio, then an adjustment to the curves is necessary. This can be illustrated by considering as before a point on a contour curve defined by parameters x, y and t. If the setting is increased to a new value kZ_R by inserting a step-down VT with ratio $1:k$ then in order to restore the relay input quantities to the same value as before, it is necessary to increase the relay point voltage by some factor with no change to the current. Thus,

$$V = E\,\frac{kZ}{Z + Z_S} = \frac{EZ'}{Z' + Z'_S}$$

$$I = \frac{E}{Z' + Z'_S} = \frac{E}{Z + Z_S}$$

from which

$$Z' = kZ \text{ and } Z'_S = -(k-1)Z + Z_S$$

Thus

$$x^1 = Z'/KZ_R = x$$

and

$$y^1 = Z'_S/kZ_R = -(k-1)x/k + y/k$$

This shows that the effect of an increase in setting due to a step-down in voltage

($k > 1$) is a reduction in S.I.R. The practical effect is to degrade the relay per-formance, and for this reason this form of adjustment is not favoured for zone-1 impedance setting adjustment. On the other hand for zone-2 adjustment, where it is preferable to express performance in terms of zone-1 setting, the range y is not significantly affected but the reach x is increased by the factor k.

9.7.5   Alternative methods of presentation

A particular form of presentation which is used to some extent shows operating time plotted against fault position x with relay-point voltage as a parameter. Obviously, a family of such curves can be derived from the superposition of constant voltage contours on the timing contours (reference Fig. 9.7.4B). Such curves do not, however, represent realistic system conditions, particularly as the fault position approaches zero (corresponding to close-in faults) because in the limit ($x = 0$) this implies infinite current to maintain the constant voltage constraint, i.e. $Z_S \rightarrow 0$ as $Z_L \rightarrow 0$.

An alternative form of presentation is to define operating time as a function of fault position, x, with the current I as a parameter, although in this case the essential property of normalising the display to relay impedance is lost. This may be avoided by using the parameter IZ_R in place of I, and although this form of presentation is an improvement on the constant voltage method it does have limitations. First, the current in a faulted power system is not constant with distance to fault and secondly if a large number of curves are drawn intersection occurs and it becomes difficult to evaluate performance.

9.7.6   Steady-state performance presentation

(*a*)   *Measuring accuracy:* Other methods of presenting relay performance are possible based on those described above in Section 9.7.5. For example, it is possible to plot relay accuracy x as a function of current or voltage. In this case, the per-unit impedance can be plotted on a linear scale and the current or voltage on a log scale as illustrated by Figs. 9.7.6A and B. The most useful of these two methods is that using current because the minimum pick-up current is easily obtained, although in practice it is preferable to use IZ_R so that the curves are made more general, i.e. they are normalised to the relay setting impedance. Such graphs provide the most convenient method for plotting the results of steady-state tests and enable characteristics of relays to be compared and assessed quickly.

(*b*)   *Polar presentation:* Up to now the polar characteristic has been used simply as a convenient means of distinguishing between the many types of comparator operating characteristic and indeed, in practice, this is how it is normally viewed. It is important to recognise, however, that although the polar characteristic should ideally have a constant shape (except in special cases such as the crosspolarised mho) over the full operating range, it is necessary in practice to define the shape for particular system conditions such as constant S.I.R., constant voltage or constant current, these being the important influencing factors.

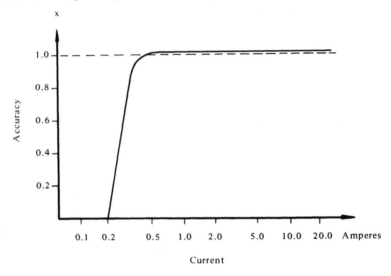

Fig. 9.7.6A   *Measuring accuracy as a function of current*

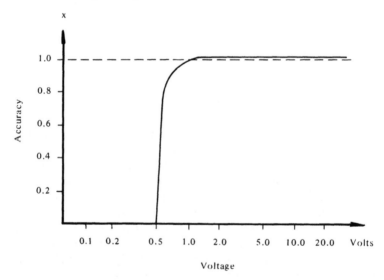

Fig. 9.7.6B   *Measuring accuracy as a function of voltage*

The preferred form of presentation would be a polar curve plotted for a given value of source impedance or S.I.R. because, as shown earlier, this represents a practical system condition at any point in time. However, unless the current and voltage inputs to the relay are being derived from a model power system so that source impedance can be maintained constant and the complex impedance to the fault varied, it is somewhat difficult to establish test conditions to simulate constant S.I.R.

An alternative presentation is the constant voltage polar curve in which the voltage input to the relay is at a constant value and the current together with its phase relationship with the voltage is varied. Obviously this is not a realistic form of characteristic because it is equivalent to a system in which the fault impedance varies with the system source impedance in order to maintain a constant relaying voltage. Such a curve is, however, useful when plotted at nominal voltage because it then gives a true measure of the relay boundary and hence the discriminative properties under steady-state system conditions. The main disadvantage with

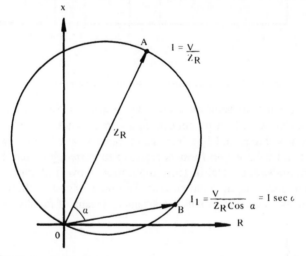

Fig. 9.7.6C *Constant voltage polar characteristic*

constant voltage polar curves is in the practical derivation. This is illustrated by Fig. 9.7.6C for a mho characteristic in which a point B on the boundary requires current I_1 for marginal operation which is related to current I by $I_1 = I \sec \alpha$. Thus, for conditions corresponding to nominal voltage where I is large and where α approaches 90°, the value for I_1 can become very large.

Because of the above problems, the polar characteristic is plotted normally for constant current or more usually constant IZ_R $(= E/(x+y))$ from eqn. 1 of Section 9.7.2. Obviously, low values for IZ_R correspond very nearly to constant S.I.R. as illustrated by the constant IZ_R curves which have been superimposed on the contour curves in Fig. 9.7.6D.

9.7.7 Dynamic polar characteristics

Reference was made above to the ideal polar presentation which would be plotted

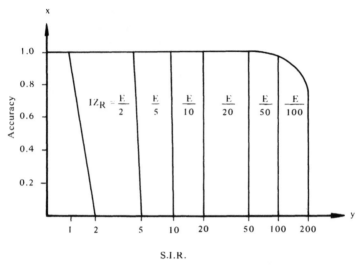

Fig. 9.7.6D   *Constant IZ_R contours superimposed on accuracy — s.i.r. contours*

normally for a constant value of system impedance or S.I.R. In practice, such a characteristic can be derived by connecting the relay under test to a model power system in which source and line impedance can be varied and in which variable-ratio current and voltage transformers are used to correctly match the relay to the system. Point-on-wave control of fault application allows full control over the d.c. transient components in the fault current. This form of system simulation has been in use for many years and is still a convenient means of rapidly evaluating relay

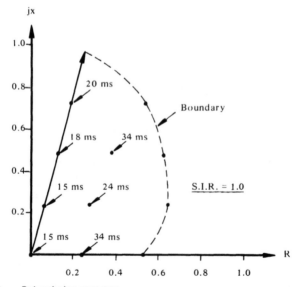

Fig. 9.7.7A   *Polar timing contours*

performance for relatively simple system configurations.

To obtain a dynamic polar characteristic, the test-bench source impedance is selected and maintained constant and the relay matched to the nominal line impedance via the C.T.s and V.T.s. The line impedance is then reduced to a known value and fault resistance added until marginal operation (corresponding to the operating boundary) of the relay is obtained. This is repeated for other values of line impedance down to zero until the boundary is well defined. Points within the boundary may be selected and speed of operation determined so as to give timing contours as illustrated by Fig. 9.7.7A. Obviously the polar timing curves, including the boundary, can be determined only for positive values of fault resistance relative to the line impedance so that only the first quadrant to the right of the line impedance is covered. However, the main feature of this form of presentation is that it automatically provides the variation in boundary due to variation of source impedance referred to earlier in Section 9.4.5. As with accuracy or S.I.R. contours, the dynamic polar curve can be derived for all types of fault condition and varying dynamic conditions, i.e. varying point-on-wave to control the transient conditions.

9.8 Switched and polyphase distance protection

9.8.1 Introduction

In many countries, the development and application of distance protection has centred on multi-element schemes, i.e. schemes in which there are measuring elements appropriate to the multiplicity of fault types. Thus, zone-1 measurement has associated with it three phase-fault elements and three earth-fault elements and zone-2 measurement may be by six separate elements or by switching the setting of zone-1 elements. There is usually at least one other zone based for example on six offset impedance elements. Such schemes have evolved over many years and are typical of the practice associated with important transmission circuits where duplication of measuring elements in the manner described is easily justified on the grounds of performance and reliability. Usually the elements associated, for example, with zone-1 are identical (assuming the same polar characteristic for earth and phase faults) with only the inputs being derived according to the type of fault being covered. On this basis it is possible to effect some economy by reducing the number of elements, for example, to a single element for zone-1 measurement (switched distance protection), or by providing an input signal arrangement which accommodates all fault types (polyphase distance protection). Such relays, having fewer components and being somewhat cheaper than full schemes, are widely applied in some countries particularly at medium and low voltage levels. In many respects, however, they are much more difficult to apply than full schemes.

Switched distance protection, as the name implies, uses switching networks to ensure that the measuring element(s) is connected to the correct phase under fault conditions. This action requires that the switching network be controlled by fault detectors which in practice may be responsive to current, a combination of current and voltage or indeed to impedance.

Polyphase distance protection does not use special switching or phase-selective networks but makes use of a special combination of relaying signals applied usually to one or more comparators. Although somewhat similar in its physical concept, the resulting characteristic is inevitably more complex, the main problem being the achievement of discriminative impedance coverage for all fault types. For this reason, polyphase relays are not as widely used as switched distance relays.

9.8.2   Switched distance protection

(a)   Principles

The application of switched distance schemes has been limited mainly to medium-voltage lines largely because of their performance relative to full schemes. One important aspect of performance is, of course, the overall operating time which is influenced by the need to first identify the faulty phase(s) and then switch the measuring element(s) to the correct phase(s). Thus, for zone-1 measurement there are three functions (starting, phase selection and measurement) involved in the measuring process compared with only one (zone-1 measurement) in a full scheme. This is illustrated by Fig. 9.8.2A which shows a simple schematic arrangement for a scheme with instantaneous overcurrent starting elements. Under fault conditions, say R-E fault, the red phase starting element operates and energises relay followers in the switching networks such that the red phase current and voltage, together with any necessary polarising voltage, are applied to the main measuring element.

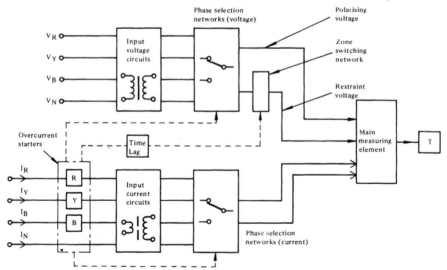

Fig. 9.8.2A   *Switched distance protection (overcurrent starting)*

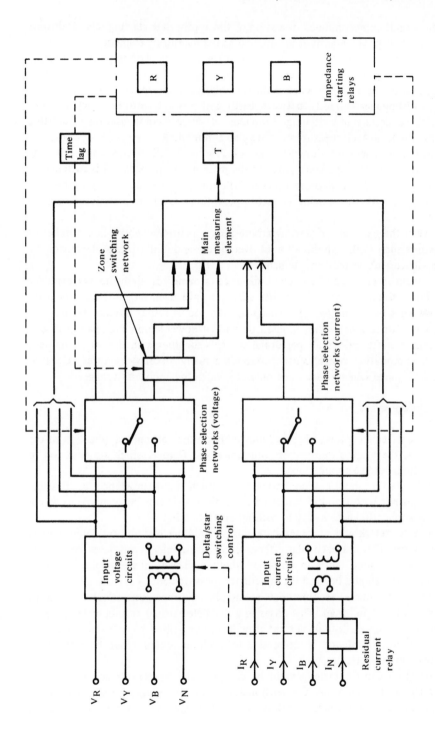

Fig. 9.8.2B *Switched distance protection (impedance starting)*

For phase faults, the combined operation of the appropriate starting relay followers will ensure that phase quantities are applied to the measuring element.

Fig. 9.8.2B shows a schematic arrangement for a scheme with impedance starting elements. The principle of operation is as described above, although correct energisation of the three impedance starting relays is somewhat complicated by the need to distinguish correctly between earth and phase quantities which requires star/delta or delta/star switching. Problems of correct compensation of starting elements for earth-fault measurement are also common.

Both the arrangements shown in Fig. 9.8.2A and 9.8.2B show a zone-switching feature whereby the impedance reach of the main measuring element is switched to a higher setting after a definite time-lag. The number of such zones may be two, three or four with the starting elements providing a final nondirectional back-up feature.

Between the extremes of starting arrangements mentioned above lie a number of alternatives such as the combination of overcurrent and undervoltage starters or the voltage-dependent overcurrent element.

The main point about switched distance schemes is that the performance limitations associated with the earlier schemes have been largely overcome with improvements in technology. In particular, the use of semiconductor techniques, reed-relay switching networks, modern designs of power supply etc., have afforded significant benefits in overall performance. This together with the use of modern rack accommodation which can accommodate many auxiliary features as well as the basic distance scheme have done much to widen the application range.

(b) Starting relays

The choice of starting relays is based inevitably on the particular application of the scheme but there are a number of requirements as follows which are common to most medium-voltage applications:

  (i) The starter should accommodate as much fault resistance as possible consistent with stability under load conditions. This is particularly so where the application is to a system without an earth-wire.

 (ii) The starter should accommodate both solidly earthed and insulated neutral systems. This has particular effects on the phase selection process for double-phase-to-ground faults in terms of whether phase-to-phase or phase-to-earth quantities should be used for impedance measurements.

(iii) The starter should accommodate sound-phase currents, under fault conditions, in excess of full load. This arises in a multiple-earthed system where proportions of the earth-fault current flow in the unfaulted phase(s) such that the apparent impedance is decreased to such an extent that correct phase selection is difficult.

 (iv) An (impedance) starter should be compensated to measure correctly under earth-fault conditions. Conventional methods of applying earth-fault compensation to impedance starters can result in sound-phase starters operating for very heavy earth-fault conditions.

Overcurrent relays: The simplest form of phase selector uses instantaneous overcurrent elements and is used mainly with the cheaper schemes associated with lower voltage lines. The main factors influencing the current setting of the starters are:

(1) the maximum load current: the drop-off value of the relay must be such that it will reset when maximum load current is flowing (this will be above normal full-load)

(2) the maximum sound-phase currents for phase-earth fault conditions in a multiple-earthed system; the sound-phase elements should not respond

(3) available fault current in the region where the starters are used to provide a (fourth zone) back-up function.

Because there are obvious limitations to the performance of schemes with only overcurrent starting it is important to have some means of identifying the application range relative to the main impedance-measuring function. This is achieved in a simple manner by the presentation derived in Section 9.7.6, Fig. 9.7.6D. Usually the setting of the starters is constrained by the above considerations so as to significantly limit the overall operating range (S.I.R.) and minimum fault MVA conditions, especially for phase-earth faults which are often not covered. Under such conditions the overcurrent starters may be combined with undervoltage starters.

Undervoltage relays: Undervoltage starting is sometimes used together with overcurrent starting, particularly on resistance-earthed systems. The starters take the form of three phase-to-neutral-connected undervoltage elements which may for some applications be controlled by a residual current element. On solidly earthed networks, the undervoltage setting is constrained by the need for reliable operation for remote faults (undetected by the overcurrent feature) while maintaining stability of the sound-phase elements during close-in earth-fault conditions.

Impedance relays: For applications where overcurrent starting relays will limit the fault coverage there is a growing tendency now to use impedance starting relays rather than combine overcurrent and undervoltage elements because the latter method provides little if any cost benefit. Various types of impedance relays have been used as described below, but perhaps the most common up to now has been the plain impedance having a circular characteristic centred on the origin of the *X-R* plane.

It is common practice to use three impedance starting elements each of which is supplied with phase-phase voltage and phase current under quiescent conditions, the voltage circuits being switched phase-to-neutral under earth-fault conditions by means of a residual-current relay. Although this form of starting significantly improves fault coverage, compared with overcurrent starting, there are fundamental problems to overcome, notably:

(i)   The use of phase current under phase-fault conditions requires that only half the normal phase-phase voltage be used as a measuring quantity so as to ensure correct measurement of positive-sequence line impedance.

(ii)   Under three-phase fault conditions, which is measured as a phase-fault, both phase-angle and magnitude errors result from the use of phase current and these have to be compensated for.

(iii)   With the delta-star switching of voltage inputs referred to above, a two-phase-earth fault is treated as a phase-earth fault when the residual current is sufficient to energise the residual current starter. Where the residual current is small, a phase-fault measurement takes place but in either case measuring errors are unavoidable.

(iv)   To compensate the impedance starters for correct earth-fault measurement, using, for example, residual current compensation, would in many applications result in incorrect operation of sound-phase starters for close-in earth-fault conditions. For this reason such compensation has usually been omitted from impedance starters with the effect that the phase-fault impedance reach is different to the earth-fault impedance reach leading possibly to application difficulties.

This particular problem may be overcome quite simply by effectively compensating the earth-fault measurement for forward faults (along the protected section) and omitting compensation for reverse faults. This is achieved, for example, for a phase comparator by using inputs:

$$S_1 = (I + k_e I)Z_R - V \qquad \text{where } k_e = \tfrac{1}{3}(\frac{Z_0}{Z_1} - 1)$$
$$S_2 = IZ_R + V$$

It can be shown that sound-phase elements are stable for close-in faults and that the polar characteristic for phase faults is centred on the origin and the characteristic for earth faults is offset. The forward reach, however, is identical for phase and earth faults.

Plain impedance starters cover the majority of applications but difficulties may sometimes be encountered on long lines where load encroachment may be a problem. This may be overcome to some extent by using offset impedance starting relays although the benefits may not be too significant for large lagging loads. Stability of sound-phase elements under earth-fault conditions is, however, increased by virtue of the reduced reverse reach of the starter and this can be further enhanced by the technique described above, i.e. applying compensation only for impedance measurement in front of the relay. The fundamental problem described in the previous Section, associated with the use of phase current for phase-fault detection, apply also to offset impedance relays.

An alternative to both the plain and offset impedance relays is the voltage-dependent impedance relay, the characteristic of which is essentially a circle in the complex impedance plane, having a diameter which increases nonlinearly as the voltage decreases. A typical voltage-current characteristic is shown in Fig. 9.8.2C,

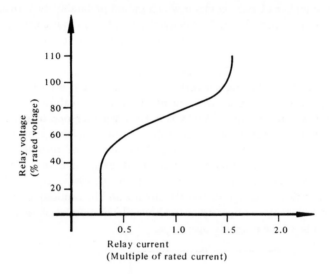

Fig. 9.8.2C *Voltage-dependent impedance characteristic*

and if the shape of this characteristic is correct then the relay impedance boundary tends to follow the load curve.

One particular reason why this type of relay has been applied is the fact that delta-star switching of voltage inputs is unnecessary for phase and earth-fault coverage, although it is still required for the main measuring element. This is perhaps a marginal benefit now that high-speed residual current relays can be designed and used with high-speed reed relays.

(*c*) *Phase selection network:* As stated earlier there are three main functions in switched distance schemes which may be defined as fault detection, phase selection and impedance measurement. Each of the fault detection or starting elements must energise follower relays which are then connected to form the phase selection network, the function of which is to switch the correct inputs under fault conditions to the main measuring elements. There is an obvious requirement to switch both current and voltage signals and where a crosspolarised element is used, a third switching network will be required for the sound-phase signal switching.

The main requirements of any phase-selection network can be classed broadly as follows:

(i) Contact performance: the most important aspects of the relay contact performance are the VA capability, contact resistance and voltage withstand across open contacts. The relay must be able to handle the power levels applied to the measuring comparator which should not be so small as to present problems due to contact resistance. Excessive voltages must be limited to within the voltage rating of the contacts.

(ii) Contact configuration: the contact configuration should be such as to minimise the number of relay contacts which should preferably be normally open contacts rather than change-over contacts. This latter requirement is to avoid problems due to sticking contacts. An example of current and voltage switching, based on changeover contacts and controlled by impedance starters is shown in Fig. 9.8.2D. The voltage circuit switching shows two sets of input terminals to the switching network, the polarity of one set being controlled by the delta-star switching arrangement. A particular feature of this arrangement is that the selection process is correct whether one or more starters operate. For example, under phase-fault conditions involving red and yellow phases it is possible that only the red phase starter will operate if the fault current is low. On the other hand if the fault level is high then both red and yellow starters will operate. In either case the red-yellow voltage is selected.

It is possible by design of the VT circuits and incorporating special logic arrangements of starter operations to reduce the number of physical contacts, which can be of the normally open type.

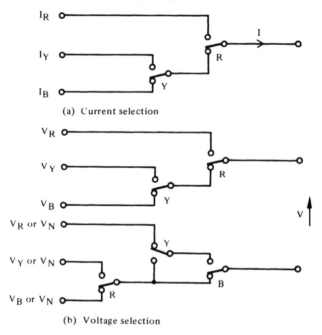

(a) Current selection

(b) Voltage selection

Fig. 9.8.2D   *Phase selection networks*

Where a switched distance relay is applied in a high-impedance or Petersen Coil earthed system, the contact configuration is often prescribed because of the problem of 'cross-country' faults, a term used to describe simultaneous earth faults which occur on different line sections and on differing phases.

For such conditions it is usual to use phase-to-neutral voltage and phase current for measurement and also to have a facility for phase-preference switching, i.e. faults involving R and Y phases, preference may be given to clearance of phase R and for faults involving Y and B phases, preference may be given to phase B. Thus the ascribed phase preference is R before Y before B, but usually this can be changed to any of the other five phase preference arrangements. Such a change in preference may be necessary when networks are electrically interconnected but the phase designation of conductors is not the same. Once one fault is cleared the other fault remains on the system but is extinguished by the Petersen Coil.

(iii) Speed of operation: developments in relay technology have done much to improve this aspect of performance. The most significant improvement has been the transition from conventional relays (e.g. attracted armature types) to encapsulated reed relays and certain types of semiconductor switching device.

Phase selection based on conventional electromechanical relays including modern high-speed designs, result in a switching time of 5-10 ms, whereas reed relay switching amounts to between 1 and 2 ms.

Wherever possible, the above features should be designed into a phase-selection network to ensure a high-speed performance coupled with high integrity. The number of switching networks required will, of course, depend on the type as well as the number of main impedance measuring elements.

(d)   Impedance-measuring arrangements

The main impedance-measuring arrangement can vary from a single element of the mho type (or indeed of a more complex shape) to two elements, one for phase-fault measurement and one for earth faults. This latter, dual element, arrangement would invariably use elements having different characteristics, e.g. a mho-element for phase faults and a reactance element, or perhaps only a residual current element, for earth faults. Phase selection networks are, of course, more complex for dual element schemes.

The single-element arrangement is still the most common, although there is now a wider range of characteristics used ranging from the traditional circular to the quadrilateral shapes. Two or more zones of measurement are invariably provided by the single element with the necessary zone switching being initiated by the appropriate time-lag circuits which in turn are initiated by the fault detectors.

In the case of contemporary schemes based on semiconductor technology there is obviously a requirement for power supplies which are derived often from a d.c./d.c. convertor. Quiescent drain on the battery can be minimised, in some cases to zero, by using the starters to switch the supply to the main circuits; the starters themselves may be self-energised.

9.8.3   Polyphase distance protection

As stated in Section 9.8.1, the polyphase distance relay is based on a special combination of relaying signals which are applied to one or more comparators. A true polyphase comparator in which the input signals can accommodate all fault types, including three-phase, has not yet been realised. Practical schemes have been based sometimes on separate measuring systems for earth and phase faults, with, in most cases, a third relay for three-phase faults. The earliest forms of polyphase relay used the 4-pole induction cup element, the torque of which is given by:

$$S_1 S_2 \sin \phi$$

in which S_1 and S_2 are the input signals and ϕ is the angle between them. Thus the two signals should either be in phase or antiphase at the boundary of operation. For phase-fault coverage the input signals would be of the form

$$S_1 = (I_R - I_Y) Z_R - V_{RY}$$
$$S_2 = (I_R - I_B) Z_R - V_{RB}$$

for directional impedance measurement. It is possible also to use these signals applied to a static $\cos \phi$ comparator provided a $90°$ phase shift is introduced between the signals.

The above signal arrangement will give correct impedance measurement for all interphase faults but not for three-phase faults. The analysis of the characteristic for various fault types is complex because, for any phase-fault condition, one or both inputs has a proportion of sound-phase voltage applied. However, it can be shown using the simplified analysis in Section 9.4.6 that the polar characteristic always passes through the extremity of vector Z_R for all phase faults but that the orientation of the circle (which encloses the origin for forward faults) with respect to Z_R varies according to the particular phases involved. For phase faults behind the relay location the relay characteristic lies wholly in the third and fourth quadrants of the Argand diagram.

An alternative configuration of inputs is based on phase current and phase-neutral voltage and affords protection against both phase and earth faults. Zero-sequence current compensation is used to ensure correct operation for phase-earth faults. Thus the inputs may typically be derived from phases R and Y as follows:

$$S_1 = (I_R + k_e I_r) Z_R - V_R$$
$$S_2 = (I_Y + k_e I_r) Z_R - V_Y$$

Such an arrangement will not accommodate earth faults on the blue phase for which an additional single-phase element is required; this additional element also covers three-phase faults.

The above has considered only representative polyphase relaying arrangements based on phase comparator principles. It is possible to use amplitude comparison although generally a practical arrangement is based on three single-phase comparators, the outputs of which are paralleled. This often affords better coverage than the phase comparator and has been most successfully implemented using rectifier-bridge comparator schemes. Each comparator is arranged to operate when

$$|(I - k_e I_r) Z_R| > |V|$$

where I and V are phase current and phase-neutral voltage of the appropriate phase. Such an arrangement covers all phase-faults including three-phase and all earth faults but requires effectively three comparators.

In general polyphase relaying has not been widely applied principally because of the difficulty in predicting the performance for all conditions and because, for most schemes, phase selection and phase flagging are not possible.

9.9 Distance protection schemes based on information links

9.9.1 General

Distance protection, based as it is on impedance measurement using voltage and current signals at a single location on the feeder, has been shown to be a very flexible protection having a wide range of possible operating characteristics. In its basic form, however, external faults are distinguished from internal faults only at the expense of delayed operation for those internal faults which are beyond the zone-1 setting. It is to overcome this particular disadvantage and/or to provide co-ordinated tripping between line ends for autoreclosing or on teed and tapped circuits that distance protection is usually used with information links over which simple command information is conveyed. Thus a system may use one-way or two-way transmission of basic two-state information which in turn is the result of a measuring process such as, for example, the operation of a zone-1 measuring element. By this means it is possible to achieve high-speed and co-ordinated clearance of faults along the whole feeder length, although overall performance has to include that of the communications channel particularly in terms of availability, effects of interference and so on. The principles involved in transmitting information over the various types of information link (pilot-wire, power-line-carrier, etc.) are covered in Chapter 7.

The methods of using information links with distance protection can be categorised broadly as those based on a command 'to trip' and those based on a command 'not to trip'. Tripping schemes can be further subdivided according to whether the received signal is used for direct tripping or whether tripping is dependent on the conditions of relays at the receiving end as well as on the received signal. In practice, there are many types of scheme particularly now with solid-state designs which allow schemes that would have been too complex and costly using traditional electromechanical protection. The following considers the basic operating principles of the more important schemes of distance protection based on information links. Various aspects of the application of these schemes are covered in Chapters 10 and 17.

9.9.2   Tripping schemes

There are three basic arrangements of protection based on a command 'to trip' and these are direct intertripping, permissive underreach and permissive overreach. Each of these is described in the following Sections.

9.9.2.1   Direct intertrip

The principle of direct intertripping applied to distance protection is illustrated by Fig. 9.9.2A, in which operation of the stage-1 relays at either end will initiate tripping at that end and also result in transmission of information to the remote end, the receipt of this information being sufficient to initiate tripping without any additional control by remote-end relays. It is apparent that the type of communication channel used for direct intertripping is a significant factor in the overall performance of the scheme particularly in terms of the effects of interference present in the system at the time of the fault. In this respect, both the type of information link and the manner in which information is communicated are critical factors, the object being to achieve correct tripping in the presence of interference and minimise incorrect tripping, the probability of which increases as the speed of the communication channel increases.

Any of the information links described in Chapter 7 may be used, although pilot wire and power-line-carrier have been the most widely applied. Methods of communication can vary significantly, particularly over pilot wire links, with schemes ranging from high-speed d.c. communication for short distances to slow-speed d.c. communication for longer distances where induced voltages in the pilots may be a problem. More sophisiticated methods are used over rented telephone pilot wires/ channels where it is usually necessary to provide coding of the transmitted information, the coding ranging from relatively simple frequency-shift keying to more complex coded pulse trains. Generally, the more complex the code, the slower will be the end-to-end tripping function. Frequency-shift schemes are the most widely applied in p.l.c. systems of communications.

Of all the tripping schemes of protection, which up to now have been preferred to blocking schemes, direct intertripping has had only limited usage being applied mainly to transformer feeders where a circuit breaker is not associated with the transformer. In some cases it has been used as a back-up protection because of its independence of relay controls.

9.9.2.2   Permissive intertrip — underreaching schemes: Permissive intertrip protection of the underreaching type exists in various forms. The basic principle,

as described under direct intertripping, is still to convey zone-1 operation at one end or the other of the protected circuit over an information link in order to accelerate an otherwise zone-2 tripping time towards zone-1 time. In the case of permissive schemes, the tripping action of the received signal is made dependent on fault detecting relays, rather than allowing direct tripping as in Fig. 9.9.2A, and such relays can of course be zone-2 or zone-3 distance relays. There are two basic arrangements which can be classed broadly as:

(*a*)   schemes in which the controlling relay operates independently of the received intertrip signal, and

(*b*)   accelerated schemes in which the controlling relay operates only on receipt of the intertrip signal.

These schemes are considered in more detail in the following.

Permissive underreach with independent control: An example of this type of 18-element protection is shown in Fig. 9.9.2B. The essential difference between this and the accelerated scheme below is that the controlling relays (which are shown as independent zone-2 elements but could equally be the zone-3 elements) can measure and operate independently of the received signal. Thus (instantaneous) operation of the zone-2 relay *and* operation of the receive relay will result in tripping.

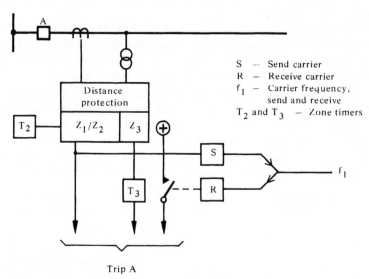

Trip A

Fig. 9.9.2A   *Direct intertripping*

Accelerated distance protection: In this form, the distance protection is usually of the 12-element type in which 6 elements are used for zone-1 measurement and are switched to measure as zone-2 elements after a time-lag; 6 elements are used for zone-3 elements. Thus for a fault close to one end of a protected circuit the zone-1 relays at that end will operate and cause transmission of information to the remote

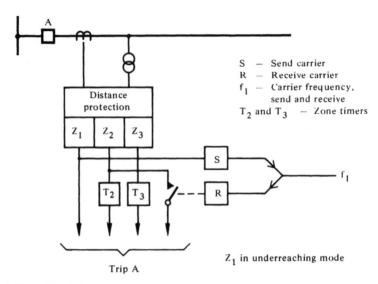

S   —   Send carrier
R   —   Receive carrier
f_1 —   Carrier frequency,
             send and receive
T_2 and T_3 —   Zone timers

Z_1 in underreaching mode

Trip A

Fig. 9.9.2B   *Permissive underreach — independent control*

end. When the receive relay operates at the remote end, the setting of the zone-1 relays is switched to the zone-2 (overreaching) setting and the zone-2 time-lag is bypassed to allow tripping.

It will be apparent from the above that, with permissive schemes, failure of the information link does not prevent correct operation of the basic 12- or 18-element distance protection insofar as the time-distance characteristic would be retained with delayed tripping for internal faults beyond the zone-1 setting.

The main advantage of permissive intertripping over direct intertripping is that the dependence of tripping on both received command and local fault detecting relays reduces the risk of unwanted tripping in the presence of noise. The risk is not, however, entirely removed because the controlling relays are necessarily of the overreaching type in which the setting extends beyond the remote terminal and there is therefore a finite risk of instability for external faults within the reach of

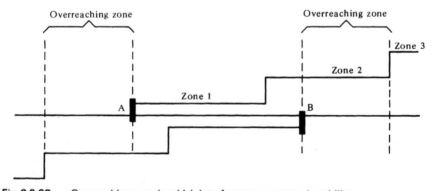

Fig. 9.9.2C   *Overreaching zone in which interference may cause instability*

the fault detectors if the receive relay responds to noise generated by the fault and/or the operation of circuit breakers. In this respect, zone-2 control relays are preferred to zone-3 because of the reduced overreaching zone (Fig. 9.9.2C). Also, if the benefits of improved speed of operation are to be retained, there is a need to use only the simplest of coding of the transmitted carrier so that, for example, single frequency-shift methods would be preferred for a p.l.c. system. Plain uncoded systems have been used but only where the information link is not subject to high levels of interference.

9.9.2.3   Permissive intertrip – overreaching systems: A permissive overreaching scheme differs from the above in that the initiation of a tripping command is by distance relays which are set to reach beyond the end of the protected circuit. Such overreaching elements may be zone-1 relays with extended reach or separate instantaneous zone-2 relays. A typical arrangement based on zone-1 is illustrated in Fig. 9.9.2D. With the zone-1 reach extending beyond the circuit end, it is necessary that tripping be dependent on both the operation of a zone-1 relay and the receipt of a tripping command from the remote end. Thus, tripping at each end depends on the operation of zone-1 relays at both ends. This means, of course, that if for any reason, such as a low fault level, the tripping at one end is delayed then tripping at

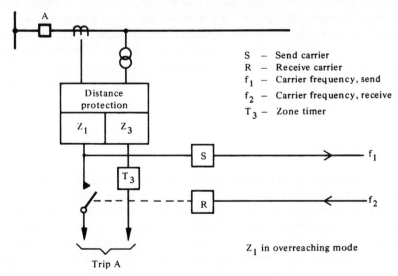

Fig. 9.9.2D   *Permissive overreach*

the other end, which may be a high fault level, will also be delayed. If in the extreme case, a circuit breaker at one end is open, the tripping of the other end will not take place and means must be found to make tripping independent of the open end.

The comments made earlier in Section 9.9.2.2 concerning the possibility of instability for external faults in the overreach zone due to system-generated noise apply also to overreaching schemes. The effect of noise during internal faults is important only if the true carrier signal can be suppressed by the noise, in which case there will be a possibility of delayed operation.

Other aspects of performance which are disadvantageous relative to under-reaching schemes are as follows:

(a)   Because a tripping command is initiated by an overreaching element, it is necessary for transmitted and received commands to be of different frequencies so that local transmission cannot operate the local receiver thereby causing a trip condition for external faults.

(b)   For the scheme in Fig. 9.2.2D, it is not possible to have discriminative distance protection in the absence of the information link (due for example to failure) unless special measures are taken. Thus, because of its critical role, the link may in practice be continuously supervised so that in the event of failure it can be arranged that the distance protection is modified to a normal stepped time-distance scheme. This action is not necessary if separate zone-1 and zone-2 relays are incorporated in the protection. Also the inclusion of a separate underreaching (zone-1) relay provides high-speed tripping for in-zone faults independent of the remote end.

Despite the many apparent disadvantages in comparison with underreaching schemes, the permissive overreach scheme is widely applied largely as a result of the greater fault coverage afforded by the overreaching relays (compared with the underreaching type) which can accommodate greater levels of extraneous fault resistance along the length of the protected circuit. Also, if the impedance of the protected circuit is very small (e.g. a cable circuit) it may not be possible to set the impedance of the distance relay low enough for an underreach scheme in which case an overreach mode may be adopted.

9.9.3   Blocking schemes

In contrast with the schemes described in Section 9.9.2, which are based on a command 'to trip', an alternative class of protection is based on a command 'not to trip'. Such protections are commonly known as blocking schemes and the principle is one of detecting reverse faults at a relay location, i.e. faults external to the protected circuit, and using this to initiate a command via the information link which blocks tripping at the remote end. The distance relays which initiate tripping at each end are set in an overreach mode. Thus any tendency to trip for a fault beyond the remote end will be blocked by receipt of a command from the remote end where the reverse fault has been detected. It is essential that all external faults detected by the overreaching tripping relays at one end be detected by the block initiation relays at the other end. Either end may block the other so that it is not necessary to distinguish between transmitted and received commands, i.e. a single channel may be used.

It is apparent from the above that the information link is an essential part of the blocking scheme and without it the protection would operate for external faults within the reach of the overreaching relays. However, the use of a link in a blocking mode avoids the problem of spurious tripping due to interference unless the true command signal is suppressed by the interference. For internal faults, spurious operation of the receiver due to interference may cause a short delay in tripping but this is rarely significant.

Although, in the past, blocking schemes have not been as popular as tripping schemes, largely because of their slower speed, they are now finding wider application with the advent of very high speed starting relays for initiation of the blocking command.

There are various types of blocking scheme which have been used in practice and these are considered in the following Sections.

9.9.3.1   Distance protection blocking scheme: The various arrangements of distance protection blocking are distinguished mainly by the way in which the blocking command is initiated.

Blocking initiated by nondirectional impedance relays: This arrangement, shown in Fig. 9.9.3A, uses high-speed distance relays Z_3 (offset impedance or plain impedance) for initiation of the blocking command. Interruption of the command is by the zone-1 relay which is seen to be set beyond the end of the protected circuit. For an external fault at F1, within the reach of the local block-initiation relay Z_{3A} and the remote tripping relay, Z_{1B} a block command is transmitted from A to B. A block command is also initiated at end B by Z_{3B} but is removed by the Z_{1B} tripping relay. Thus both ends stabilise, but it is important to note that the tripping of either end must be delayed to allow receipt of a blocking command for external faults. For this reason, a short time-lag is introduced into the tripping circuit as shown. For an internal fault F2 a blocking command is initiated and then removed at both ends A and B which are therefore allowed to trip.

As far as the information link is concerned, all types may be used, but because it plays an essential part in the correct operation of the scheme, particularly for external faults, supervision features are beneficial and indeed these may be arranged to change the settings of zone-1 relays to an underreaching mode in the event of a failure of the information link.

Blocking initiated by current relays: An alternative to the above arrangement is one which uses current-operated (nondirectional) relays in place of distance relays for initiation of the blocking command. In practice, it is necessary to ensure that for marginal fault conditions at one end of the protected circuit, a blocking signal is always transmitted for external faults within the reach of remote end relays and also

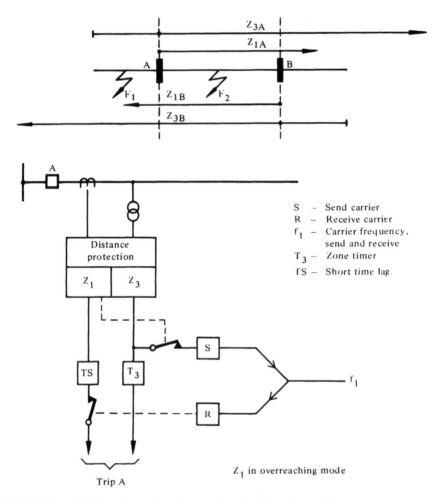

Fig. 9.9.3A   *Blocking scheme based on non-directional impedance relays*

that the blocking command is always removed for marginal internal faults. This co-
ordination of sensitivities is somewhat more difficult to ensure when the blocking
and tripping relays are of different types (current and impedance) and for this
reason a two-stage current starting arrangement is preferred. An arrangement which
illustrates the point is shown in Fig. 9.9.3B. The low-set (L.S.) current relay
initiates the blocking command at either end and the overreaching zone-1 relay
removes the command. The zone-1 relay also initiates tripping via the short time-lag
element and a high-set (H.S.) current relay. It must be arranged that the overreaching distance relays detect all internal faults and remove any locally transmitted
blocking signal so that in this respect the distance relay should be more sensitive
than the L.S. current relay. For marginal external faults there is a possibility that
the overreaching distance relay may operate and the L.S. current relay may not, in

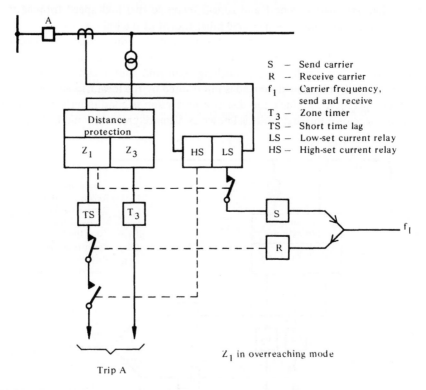

S — Send carrier
R — Receive carrier
f_1 — Carrier frequency, send and receive
T_3 — Zone timer
TS — Short time lag
LS — Low-set current relay
HS — High-set current relay

Z_1 in overreaching mode

Trip A

Fig. 9.9.3B   *Blocking scheme based on two-stage current relays*

which case stability is maintained because tripping at both ends is controlled by the H.S. current relays.

Blocking initiated by directional impedance relays: The problem of co-ordination of sensitivities of the tripping and block-initiation relays exists to some extent with schemes based on nondirectional impedance relays and to a greater extent with schemes using single-stage current-operated relays. It is possible, however, to avoid the problem completely by using either two-stage current relays or directional relays for initiation of the blocking command. A scheme based on reverse directional impedance relays Z_{1R} is shown in Fig. 9.9.3C. The reach of the reverse directional relay Z_{1RA} extends beyond that of the remote tripping relay Z_{1B} and initiates a blocking command for external faults within its reach. However, a blocking command can never be initiated for internal faults and it is required only to ensure that the reverse reaching relays operate for all faults seen by the remote overreaching relays.

All of the schemes described above in Section 9.9.3.1 have shown the tripping relay as an overreaching zone-1 relay. In practice, it is possible to use distance protection having separate zone-1 and zone-2 relays so that high-speed tripping from one or both ends by zone-1 is achieved independent of the link.

Finally, it is worth noting that in some countries, continuous transmission of signals may be allowed and it is therefore possible to avoid the use of starting relays. For all external faults, the blocking command continues to be transmitted and for internal faults the command is interrupted by the local tripping relays. This mode of operation in which the distance protection is continuously blocked and then unblocked for internal faults is known as distance protection unblocking.

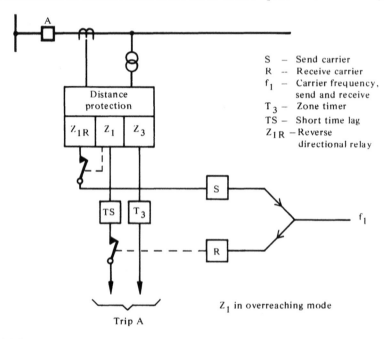

S — Send carrier
R — Receive carrier
f_1 — Carrier frequency, send and receive
T_3 — Zone timer
TS — Short time lag
Z_{1R} — Reverse directional relay

Z_1 in overreaching mode

Trip A

Fig. 9.9.3C   *Blocking scheme based on directional impedance relays*

9.9.3.2   Directional comparison: This type of protection has been widely applied and is becoming increasingly popular in modern schemes of distance protection particularly where difficult earth-fault conditions prevail. In its more conventional form, the scheme uses an arrangement of directional and nondirectional current relays as shown in Fig. 9.9.3D. The phase-fault directional relays are polarised by the relaying point voltage whereas the earth-fault relays may be polarised by either voltage or some reference current or a combination of both. Transmission of the blocking command is initiated by the nondirectional current relays which, for the reasons described in Section 9.9.3.1, are generally two-stage relays with high-set and low-set operating levels. The low-set initiates transmission of the blocking command and the high-set controls the output tripping circuit.

As in the case of the blocking scheme based on current starting, the phase-fault coverage may be unacceptably restricted by the high-set current setting, particularly where the minimum fault currents are comparable with maximum load currents. In such cases relays responding to rate-of-charge of current or, of course, distance relays, may be used. For earth-fault conditions, a much greater sensitivity is possible which makes the directional comparison particularly useful where the level of earth-fault current may be limited, due, for example, to high ground resistivity.

An alternative arrangement which is becoming more popular for difficult earth-fault conditions, uses two directional relays, connected back-to-back, i.e. one detects forward faults and the other detects reverse faults. The reverse reaching relay initiates the blocking command whilst the forward reaching relay removes the blocking and also initiates tripping in the manner described above.

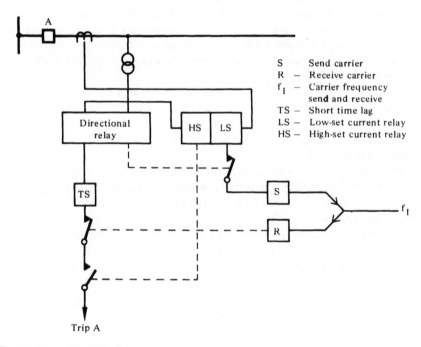

Fig. 9.9.3D   *Directional comparison scheme*

9.10   Practical considerations in the application of distance protection

9.10.1   Fault resistance

Of fundamental importance in assessing the effectiveness of the available polar characteristics of impedance-measuring schemes is the range of fault resistances likely to occur in practice. The majority of primary system faults are accompanied by a fault arc when, for example, phase conductors clash or co-ordinating gaps flash-over. The resistance of the arc is nonlinear with the most generally accepted expression for its value being due to Van Varrington, namely

$$R_f = \frac{2613L}{I^{1\cdot4}}$$

where L is the length of arc in metres and I is the current in amps.

Clearly, for a given length of arc (due, for example, to a specific co-ordinating gap length) the fault resistance increases nonlinearly as the fault MVA reduces, i.e. as the source impedance ratio increases. Generally speaking, for an earthed construction line, the fault resistance will be unlikely to exceed about 0·5 Ω which will only be significant for short lines for which indeed, fault resistance imposes an application limit. However, such a level of fault resistance does not present any major problem except perhaps if the fault occurs at or near the relay location. The performance of the relay for such faults depends to a large extent on the effectiveness of the polarising signal.

It is, of course, possible for the arc resistance to exceed the limits defined by co-ordinating gaps when, for example, a primary conductor breaks and is in direct contact with earth or where a flashover to vegetation occurs. The fault resistance can be much higher and present protection problems similar to those encountered for unearthed construction lines.

Another aspect of fault arcs which is rarely considered is the effect of the arc-voltage on the waveform of the fault transient current. The effect can be significant even for relatively low values of arc voltage but complete analysis is difficult because of the complex arc waveform. If however a relatively small arc voltage of rectangular waveform is considered the resulting transient current can be obtained by superimposing the currents resulting from the two separate voltage sources, one being the normal prospective transient current without arcing and the other being the current resulting from the replacement of the arc by an equivalent square-wave generator. An approximate evaluation of the increased rate of decay due to the presence of the arc is given by the equivalent time-constant.

$$T^1 = \frac{T}{1 + kX/R}$$

where X/R is the natural value of the circuit

k is the per-unit value of arc-voltage referred to the peak sinusoidal voltage

T is the time constant with a solid fault (i.e. no fault arc)

As an example, consider $X/R = 15$ and $k = 1/25$ from which

$$T^1 = 0\cdot625T$$

The result is, of course, approximate but serves to illustrate how the fault arc can reduce the severity of the d.c. transient associated with the fault current, this being more significant now when the X/R associated with modern plant can be extremely high.

9.10.2 Measuring errors

Accuracy of impedance measurement is of particular importance in setting the zone-1 reach for an underreaching mode, and consideration must be given to relay setting tolerance, dynamic inaccuracies reflected usually as overreach in the presence of d.c. offset transients, and the effects of c.t. and v.t. transformation errors. Good design should ensure a measuring accuracy which is maintained within ±5% to ±10% over a wide range of S.I.R. and with maximum d.c. transient content in the input signals. Even so, it is common practice to set the zone-1 reach to 80-85% of the protected feeder length for phase-fault measurement and in certain circumstances even lower for earth-fault measurement.

Earth-fault measurement is complicated by a number of factors not least of which is the uncertainty of determining accurately the zero-sequence impedance to which the relay must be set. The system impedance can vary due to changes in earth resistivity or because of mutual coupling between parallel circuits. Practical levels of mismatch between relay zero-sequence setting and system zero-sequence impedance for practical values of positive- to zero-sequence current ratio, can yield

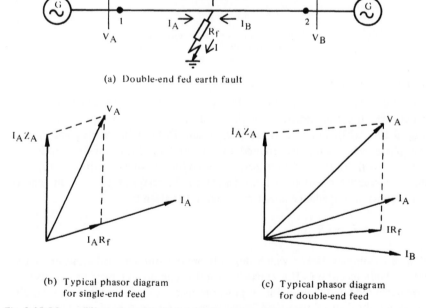

(a) Double-end fed earth fault

(b) Typical phasor diagram
for single-end feed

(c) Typical phasor diagram
for double-end feed

Fig. 9.10.2A *Effect of remote fault infeed on measuring accuracy*

measuring errors of typically ±5% on solidly earthed systems. This could possibly increase to about ±10% for lines without earth wires because of the greater influence of earth resistivity.

In addition to the uncertainty of matching the zero-sequence impedance of relay and system, there is the basic problem of accommodating fault resistance, the value of which as seen by the relay may differ from the real value as a result of remote fault infeeds. The problem is illustrated by Fig. 9.10.2A, which is simplified by assuming that source impedances at A and B are significantly larger than Z_A, Z_B

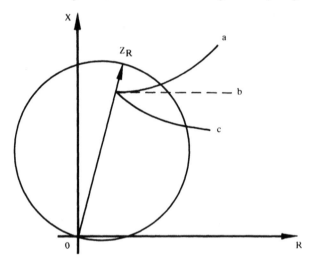

a - Power flow out of relay location
b - Zero pre-fault load
c - Power flow into relay location

Fig. 9.10.2B   *Effect of fault resistance on measured impedance*

and R_F. For a fault fed only from A the relay at location 1, supplied with V_A and I_A, would see the true resistance with the drop across R_F being cophasal with I_A (Fig. 9.10.2A(b)). For a fault fed from A and B, fault currents I_A and I_B will have a phase relationship determined by the prefault load transfer conditions (Fig. 9.10.2A(c)). Thus the drop across R_F is no longer cophasal with I_A or I_B and an apparent reactive component is presented to the relays at 1 and 2. In general terms the apparent impedance presented to relay 1 is given by:

$$Z_R = I_A + kR_f$$

where k is a complex factor which depends on the through-load transfer at the instant of fault inception. The apparent reactive component may be positive or negative depending on the direction of power transfer as illustrated by Fig. 9.10.2B, which shows the loci of measured impedance for different load conditions at a particular fault location.

9.10.3   Healthy phase relays

Multi-element distance relays theoretically fulfill the essential requirement for comprehensive fault coverage. However, the provision of discrete measuring elements for the multiplicity of fault conditions and zones of protection prescribes a very basic requirement, i.e. the distance relay should provide discriminative tripping whereby healthy phase elements should not respond incorrectly either to forward or reverse faults. Correct discrimination for forward faults is not essential except where autoreclosing sequences are involved; discrimination for reverse faults is, of course, a fundamental requirement.

The fact that faulted phase components of voltage and current will influence healthy phase elements is inevitable, since a R-Y fault, for example, produces voltage and current changes in R and Y phases which must be seen by B-R and Y-B measuring elements as well as R-E and Y-E elements. Similarly, A R-E fault produces a residual current which is seen by Y-E and B-E elements whilst the phase voltage and current are seen by B-R and R-Y. These effects are of course particularly severe for close-up faults fed from high MVA sources where it is indeed possible that a close-up earth-fault condition can give rise to operation of phase-fault elements. The converse is also true but it is the response of phase-fault elements to earth-fault conditions which is of significance in single-phase autoreclosing applications. The degree to which incorrect operation of healthy phase elements occurs can be controlled to some extent in modern relay designs particularly those using multi-input comparators. This, in addition to the use of phase-discrimination techniques and logic circuits, ensures correct functioning of autoreclose schemes.

The influence of close-up fault conditions on healthy phase elements for varying system impedance ratio can be illustrated by considering the much simplified homogeneous system condition shown in Fig. 9.10.3A. A simple close-in earth-fault is assumed with zero prefault load current, and using the expressions for phase voltage and current developed in Section 9.7.2 gives

$$V = \frac{Ex}{x+y} \text{ and } IZ_R = \frac{E}{x+y}$$

A close-in fault prescribes that x is virtually zero so that:

$$V \longrightarrow 0 \quad \text{and} \quad IZ_R \longrightarrow \frac{E}{y}$$

These conditions are illustrated by the vector diagram in Fig. 9.10.3A. The influence of these phase components of V and I on the R-Y phase-fault element (connected as a mho element with very little sound-phase polarising) is also shown

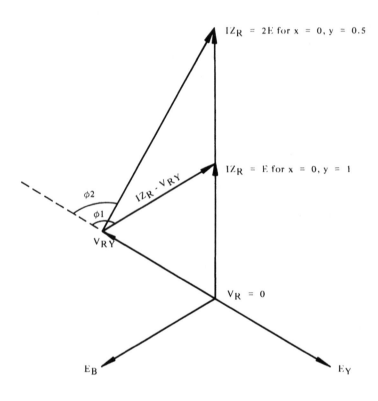

(a) Simplified system diagram

$$V = \frac{Ex}{x + y} \qquad I = \frac{E/Z_R}{x + y}$$

$IZ_R = 2E$ for $x = 0$, $y = 0.5$

$IZ_R = E$ for $x = 0$, $y = 1$

$V_R = 0$

(b) Typical phasor diagram showing relay input conditions

Fig. 9.10.3A   *Influence of close-in earth-faults on healthy phase-fault relays*

where it is seen that the phase relationship between $IZ_R - V_{RY}$ and V_{RY} can be such as to cause operation as the S.I.R. reduces. The locus of the impedance presented to the phase-fault element as the S.I.R. varies is shown in Fig. 9.10.3B. Similar loci can be drawn for B-R, B-E and Y-E elements, although the latter would tend to be well outside the operating boundary. The same kind of presentation can be used for close-in phase faults showing the effects on other elements.

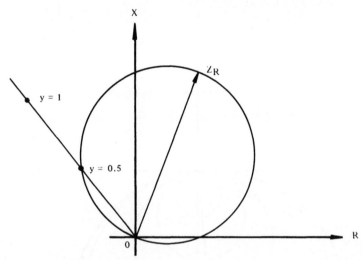

Fig. 9.10.3B   *Locus of impedance presented to phase-fault relay for terminal earth-fault and varying S.I.R.*

Of much greater significance than the above is the directional discriminating property of healthy phase elements, i.e. the effect of reverse faults on sound-phase relays. For close-up reverse faults, the fault MVA is at least limited by the protected line impedance itself coupled with the remote source impedance. The effect of reverse faults is now assessed by reversing the polar characteristic, as shown in Fig. 9.10.3C, and superimposing the impedance loci for Y-E and B-E elements. Because there is always the finite source impedance of the line itself, the healthy phase impedances do not encroach on typical zone-1 reaches so there is no danger of tripping for reverse faults. As the reach is increased however, typical of zone-2 settings, the polar characteristic can then include the apparent healthy phase impedance when the ratio of remote source impedance to protected line impedance is low. This poses a problem, particularly in permissive overreach and blocking schemes where the tripping characteristic is set greater than the protected circuit impedance. A practical limit to the reach of such characteristics would be about four times the line impedance which would satisfy most applications. This may be inadequate for teed feeders or short feeders with high remote end infeeds where special measures may have to be taken.

9.10.4   Load encroachment

One of the most fundamental aspects of applying impedance-measuring relays is the requirement for adequate discrimination against maximum circuit load conditions. The protection impedance characteristic must be such as to avoid load impedance encroachment. This is not, of course, a problem in terms of applying zone-1 and zone-2 relays but can be a significant practical problem in setting the zone-3 reach

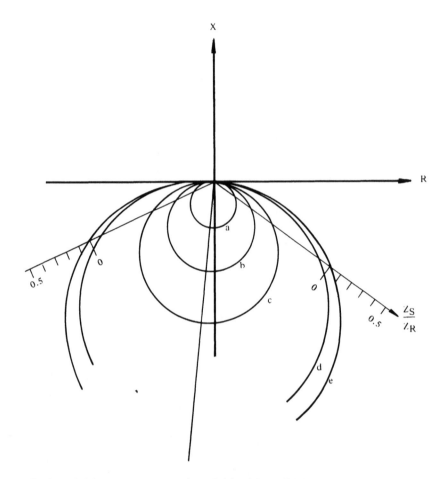

Setting of (e) equal to 6 x setting of (a) with negligible sound-phase polarizing and earth-fault compensation

$$\frac{1}{3}\left(\frac{z_O}{z_1} - 1\right) = 0.5$$

Fig. 9.10.3C   *Effect of reverse faults on sound-phase relays*

for circuits comprising very long bundle-conductor lines. Practical considerations of circuit loading must take account of maximum circuit rating as dictated by the sending-end to receiving-end busbar-voltage ratio and the corresponding angular displacement. These factors influence the boundary of load conditions in the manner shown in Fig. 9.10.4A. Here the variation of angular displacement between busbar voltages for unity voltage ratio between ends produces the load variation A, B, C. The effect of change in the ratio of busbar voltages is seen to produce the curvilinears $A'B'C'$ and $A''B''C''$ the two sets P1 and P2 being for importing and exporting conditions, respectively.

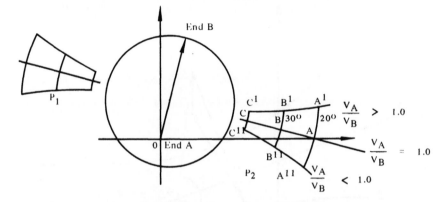

Fig. 9.10.4A   *Load conditions superimposed on impedance polar diagram*

Clearly, maximum circuit loading can present an impedance to the distance relay which encroaches on its boundary and this may lead either to a limitation on relay setting or the use of a relay whose characteristic is less effected by such considerations, i.e. elliptical or quadrilateral characteristics. A detailed assessment of load discriminating properties of a relay must include such factors as mismatch between line angle and relay angle, practical tolerances on the polar characteristic boundary, the possibility of accommodating short-time (20 min) circuit overload rating, and so on. Within these parameters, the degree of remote back-up protection afforded by the traditional circular zone-3 relays may be severely limited so that the sort of action mentioned above may be warranted.

9.10.5   Power swing encroachment

A power swing is the result of a change in angle between busbar voltages which is seen by distance relays as an apparent changing impedance which may ultimately encroach on the relaying boundary. The conditions of a power swing can be represented on a polar diagram as shown in Fig. 9.10.5A which assumes a very simple case of two generators behind source impedances Z_{S1} (corresponding to G1,0) and Z_{S2} (corresponding to G2,Q) connected by a line impedance Z_L (0,Q). The relay is assumed to be at 0. When source voltages $E1$ and $E2$ are equal in magnitude, variation of the phase angle between them produces the locus of phasor 0Z bisecting Z_T (G1, G2). When source voltages $E1$ and $E2$ are unequal, the locus of Z is a circle to one side or the other of ST, the radius depending on the ratio $E1/E2$ and the centre lying on G1G2 produced.

   When a power swing occurs, the apparent impedance may swing in from the right or from the left of the diagram, the angular velocity of the swing growing less as the centre is approached. The velocity is important insofar as the swing may

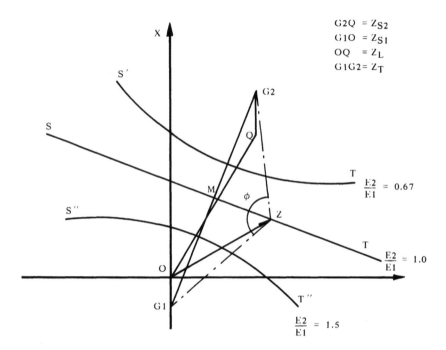

Fig. 9.10.5A   *Conditions of power swing on an impedance diagram*

decrease to zero before encroaching on primary protection characteristics so that the system will begin to swing back into phase. If, however, the velocity is so great that the swing passes through Z_T, the velocity will start to increase in the same direction, say from right to left, returning to the right via infinity at the end of one slip cycle.

Clearly, encroachment of the apparent impedance due to a power swing onto distance relay characteristics may produce indiscriminate tripping at various points in an interconnected system. This can be overcome by using 'blocking schemes' which will detect the power swing and inhibit the trip function of the main protection. It may be, however, that the power swing is such that parts of the system have fallen out of synchronism so that selective tripping is necessary in order that the main parts of the system can continue to run separately, i.e. a power swing 'tripping scheme' is required.

A blocking scheme is relatively easy to implement, requiring, for example, an additional phase-fault measuring relay, the characteristic of which is such as to completely enclose the normal zone-3 characteristic as shown in Fig. 9.10.5B. The output of the additional power swing relay P_S energises a time-lag element, the setting of which is co-ordinated with the difference in impedance settings of the zone-3 and power swing elements to provide a very basic measure of rate-of-change of impedance. This provides the basis on which a fault, which is effectively an instantaneous change in apparent impedance, is distinguished from a power swing

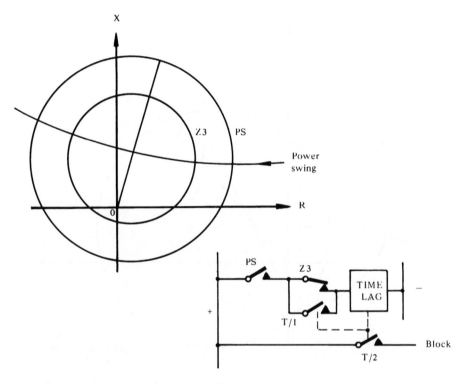

Fig. 9.10.5B   *Power swing blocking scheme*

which represents a finite rate-of-change of impedance. Thus a power swing will
first result in operation of the power swing element which then initiates the time-
lag element. If the time-lag elapses before operation of the zone-3 element the main
tripping elements of the protection are inhibited. If, however, a fault occurs within
the zone-3 boundary, the zone-3 operates before the time-lag element which is then
inhibited from operating so that performance of the primary protection is not
affected.

A tripping scheme is more difficult to implement because it is necessary to know
more about the power swing, particularly whether pole-slipping has occurred. A
means of doing this is shown in Fig. 9.10.5C, in which two offset mho characteristics A and B, are set to overlap. The characteristics are chosen so that the outer
boundary of impedance corresponds to a machine angle of about 60° on the basis
that this is the smallest angle that can be used without encroaching on the steady-
state load conditions of the system. The inner impedance boundary is chosen to
correspond with a machine angle of about 100° so that the distance protection
impedance setting is within this boundary for most practical applications. The
scheme, which can be used either in a blocking mode or in a tripping mode, is
based on the sequence of operation and reset of the two relays A and B. For
example, a complete pole slip from right to left will result in the sequence, A

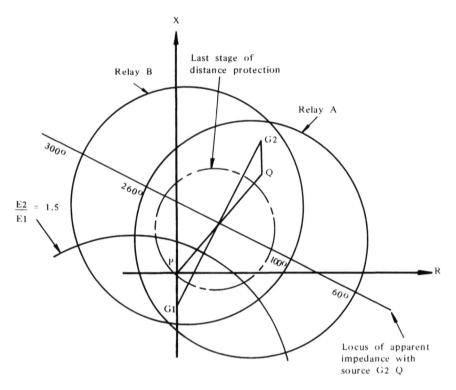

Fig. 9.10.5C   *Polar characteristics for power swing blocking and tripping scheme*

operates, B operates, A resets, B resets. A swing from right to left which moves to the 180° point and then returns will result in the sequence, A operates, B operates, B resets, A resets. Clearly such sequences identify direction and pole slipping and are sufficient to provide tripping or blocking modes of operation.

9.10.6   Line check

In Section 9.4.6, the need for crosspolarisation of mho relays was identified as a means of ensuring definite operation for close-up unbalanced faults. For the case of close-up balanced faults, all three phases-voltages collapse and, although this is a rare occurrence as a naturally occurring fault, steps must be taken to ensure relay operation. One method is to use memory circuits which comprise basically a circuit tuned to system frequency which maintains a decaying polarising voltage for a prescribed time after the system voltage collapses.

The most likely condition giving rise to a terminal 3-phase fault is the inadvertent connection of earthing-connectors when a line is energised after, say, a maintenance period. This results in virtually zero voltage at the relaying point with consequent potential failure of zone-1 and zone-2 relays so that special precautions are often taken to ensure instantaneous clearance.

A line check scheme is often used the function of which is to effectively bypass the zone-3 time-lag element for the special condition of closing onto earthing clamps. There are basically two schemes used, depending on the busbar circuit breaker configuration. The first scheme is suitable for lines controlled by one circuit breaker and is shown in Fig. 9.10.6A(*a*). It uses the circuit breaker closing action in conjunction with a timer to provide a controlled instantaneous response of the zone-3 elements. Bypassing of the zone-3 time-lag is maintained for a well defined period, typically 500ms, after removal of the circuit breaker closing command as illustrated by Fig. 9.10.6A(*b*).

The second scheme is suitable for mesh busbar arrangements where the protected line may be controlled by more than one circuit breaker. In this case, the zone-3 elements are controlled by a voltage detector operating in conjunction with two timers and the associated circuits are shown in Fig. 9.10.6B(*a*). When closing onto a terminal 3-phase fault, the voltage detector fails to respond thus leaving timer T1 unoperated and T2 operated. This allows the zone-3 timer to be bypassed, as illustrated by Fig. 9.10.6B(*b*). When closing onto a healthy system, the voltage detector responds which results in time T1 operating after about 200ms which causes the reset of timer T2 and hence the removal of the zone-3 timer bypass (Fig. 9.10.6B(*c*)). Timer T2 is required to ensure that the line-check scheme does not interfere with the normal protection function during naturally occurring fault conditions, i.e. should a fault occur which causes the voltage detector to reset then timer T1 resets and T2 starts to time-out. However, the time-lag T2 is set in excess of the zone-3 time-lag or the dead-time associated with a reclose sequence.

9.10.7   Voltage transformer supervision

One of the fundamental requirements of a distance relay is that it be supplied with signals proportional to the primary system voltage and current, one being a restraint signal and the other an operate signal. Under healthy system conditions it is essential that the input voltage to the relay be maintained since any inadvertent loss on one or more phases, with or without load current flowing in the protected line, can result in the initiation of fast tripping. Such loss of voltage may be due, for example, to accidental withdrawal of a fuse link or to a fault on the secondary v.t. wiring. Whether or not false tripping occurs further depends on other factors such as:

(*a*)   the configuration of the overall v.t. burden on the same fused circuit as the distance relay

(*b*)   the magnitude and phase of any load current.

If false tripping can occur and is considered undesirable then special high-speed detection of the v.t. fault is necessary so that the protection can be inhibited before tripping occurs. A means of detecting v.t. faults is illustrated by Fig. 9.10.7A. The scheme operates on the principle that, if residual voltage is present without residual current, then the fault must be in secondary v.t. circuits. An earth fault on the primary system would give rise to residual voltage and residual current.

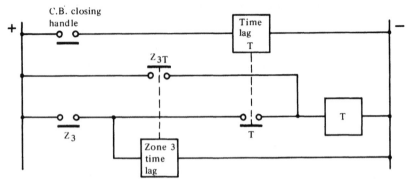

(a) Line check circuit

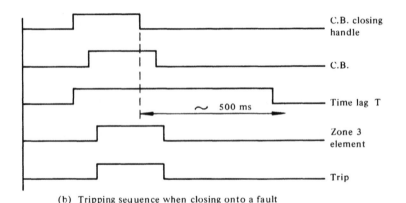

(b) Tripping sequence when closing onto a fault

Fig. 9.10.6A   *Simple d.c. line check scheme*

The residual voltage is derived via the input summation amplifier and fed to the voltage detector. The residual current detector is provided with a fast response and is arranged to inhibit the voltage detector from giving a blocking output. The output timer T can be arranged to provide delayed operation or reset. Alternatively, it can be by-passed depending on application requirements.

An alternative philosophy is to merely detect the presence of a v.t. fault and provide a local/remote alarm, the protection being allowed to operate.

9.11   Trends in distance protection development

For many years, the design of impedance-measuring protection has been based on a number of simplifying assumptions regarding the primary power system. In particular the transmission medium has traditionally been considered as a lumped

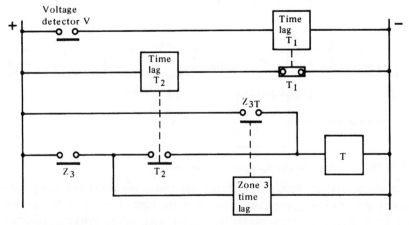

(a) Line check circuit

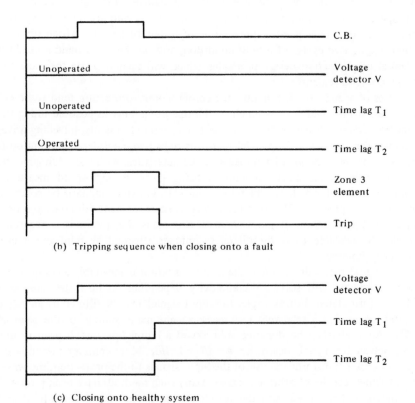

(b) Tripping sequence when closing onto a fault

(c) Closing onto healthy system

Fig. 9.10.6B *A.C. line check scheme for mesh busbar configurations*

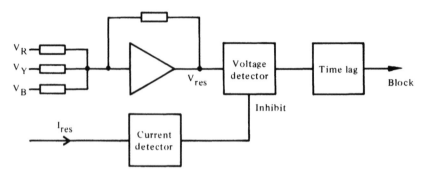

Fig. 9.10.7A   *Voltage transformer supervision scheme*

LR impedance (rather than a distributed parameter network) fed from simple generating sources having constant impedances for both steady-state and dynamic (fault) conditions. Such assumptions were valid, even under dynamic conditions, for the relatively slow electromechanical designs of protection and even for the first solid-state designs. However, the ever-increasing demand for higher speeds over wider operating conditions without sacrifice of stability and accuracy is requiring on the one hand a much greater appreciation of system behaviour, particularly over the first one or two cycles of a fault condition, and on the other hand a need for reappraisal of protection design principles which will accommodate the increased performance requirements.

In terms of speed, the design aims for contemporary protection tend to be 1/2 cycle or less maintained over as much of the operating range as possible and in the presence of extreme levels of transient, particularly d.c. transients, travelling waves and transducer transients. In parallel with this objective, considerable attention is given also to more discriminative impedance characteristics. With these fundamental requirements in mind, much thought is being given to high-speed measuring techniques, some of which are digitally-based with growing emphasis on software solutions to the problem. This trend is inevitable since the demands come at a time when the implementation of more sophisticated measuring methods is becoming more readily available using the digital techniques afforded by mini and micro-computer technologies.

Where digital methods are used, the implementation is inevitably a compromise between hardware and software complexity depending on how the measuring problem is formulated. For example, the input signals may be filtered by analogue or digital means and a relatively fast measurement made simply on the basis of $v = iZ_1$. Alternatively, the dynamic solution of a lumped-parameter primary line may be solved by digital means, i.e. $v = iR + L di/dt$. More complex methods are based on Fourier transform analysis of the input signals to the protection, the waveforms of which can be of arbitrary shape. Many such methods have been proposed and some have been tried, and the search continues for effective solutions and means of implementing such solutions. It is important to note, however, that whether the solutions are complex digital methods or relatively simple analogue-based methods the implementation is tending to favour microcomputer methods, not only because they provide a means for high-speed signal processing and data manipulation, but also because they afford a valuable self-checking facility. Both hardware and software security can be enhanced by sophisticated checking routines which add a penalty only in terms of operating-software complexity. Thus, self-diagnostic routines, containing a number of monitoring and test programmes, will verify the essential hardware and software used for protection functions. The way in which the periodic testing is done may depend on the actual protection algorithm. For example, an algorithm based on calculation of impedance can be tested without any special adaptation of input signals or 'relay' criterion. However, an algorithm based on an impedance balance point (such as a block-average comparator) may require injection of artificial signals or adaptation of load-condition signals to force operation (but not to cause a trip output).

Inevitably the opportunity will exist for microcomputer-based distance schemes to store and display fault performance and other data, e.g. operating-time, fault level, fault location possibly, and so on. Such information may be displayed locally or remotely over links which may be used also for remote adjustment of protection parameters.

Ultimately the full benefits of new types of distance protection will be enhanced when higher-speed communication channels are fully available such channels providing transmission of quantitative as well as qualitative information probably using optical links.

9.12   Bibliography

Papers

'Review of recent changes in protection requirements for the CEGB 400 and 275kV systems' by J C Whittaker
IEE Conf. Publ. 125, pp.27-33. Developments in Power System Protection.
'Polarised mho distance relay' by L M Wedepohl (Proc. IEE, 1965, **122**)
Rushton, J, and Humpage, W.D. 'Power studies for the determination of distance-protection performance' by J Rushton and W D Humpage (*ibid.*, 1972, **119**)

Books

Teleprotection CIGRE Committee Nos. 34 and 35, Report of the joint Working Group on Teleprotection (March 1979)
Protective relays; their theory and practice (Vols. I and II) by A R Van C Warrington (Chapman & Hall)

Chapter 10

Feeder protection: pilot-wire and carrier-current systems

by F.L. Hamilton, revised by L. Jackson and J. Rushton

10.1 General background and introduction

Unit protection, with its advantage of fast selective clearance, has been used for interconnections within power systems from the early stages in the development of generation, transmission and distribution. In these early stages, generation was local to the load, and distribution was the main function of the network. At this period, we find unit protection of cable interconnections an example of the first applications of the principle of unit protection. As the main network was underground, it was natural that the secondary interconnections necessary for unit protection should be by means of 'auxiliary' or 'pilot' cable laid at the same time as the primary cables required to be protected. Also, as the 'Merz-Price' or 'differential' principle was developed in this period, it was understandable that this principle should be used as the main basis of these early forms of unit feeder protection.

Since that time, unit feeder protection has developed considerably and many variations and complexities have resulted, but it is well to remember that its origin was basically simple. It should be noted also, that Great Britain has been one of the most active users of the unit principle of protection. The USA and certain English-speaking countries overseas such as South Africa and Australia have also used it widely. The problems of adequately protecting modern power systems, with their high degree of interconnection and with the need for fast clearance times have lately led Continental countries to consider adopting unit protection (compared with distance protection), and its use in these countries, therefore, is likely to increase.

The simple concept of unit protection for underground feeders using pilot wires was, by necessity, extended to overhead interconnectors as systems developed. At the same time the problem of providing economically an auxiliary channel by means of pilots directed attention to carrier techniques, which use the main primary conductors, and radio links which use aerial transmission.

10.2   Some basic concepts of unit protection for feeders

It will be appreciated that unit protection required the interchange, between the several terminations of the protected zone, of information about the local conditions existing at each termination. Feeder protection which uses information from one point only of a network is inherently non-unit, even though the protective system may be very complex in form and sophisticated in its action. An essential feature of unit protection therefore is the provision of a channel over which information can be passed between ends. The variety of unit systems of feeder protection which have been developed over the years is due to the several different methods of providing a suitable channel and the problems which each kind of channel has presented in its application.

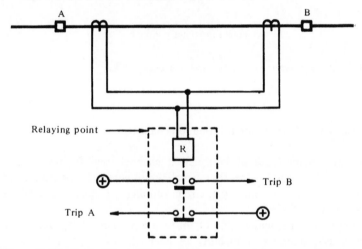

Fig. 10.2A   *Simple local differential protection with short interconnections between c.t.s*

When dealing with the zone of protection of, say, a generator or transformer, the circuit-breakers controlling the terminations of the zone are relatively close together and therefore usually require only a single 'relaying point'. The need for an information channel still exists, however, but it is relatively simple and is formed by direct interconnection of the current transformers with the single relaying point, the tripping of the associated circuit breakers being effected from that point (Fig. 10.2A).

If the terminations and their associated circuit breakers are separated by greater distances, say thousands of metres, the use of a single relaying point is no longer practicable. This means that a relaying point must be provided at each circuit breaker, and the information channel becomes more complex, each relaying point being provided with information about all the others in order to be able to assess this information together with its own and also to trip or stabilise as the case may be (Fig. 10.2B). Attenuation, distortion and delay in transmission are all important to some degree or other because significant distances are involved.

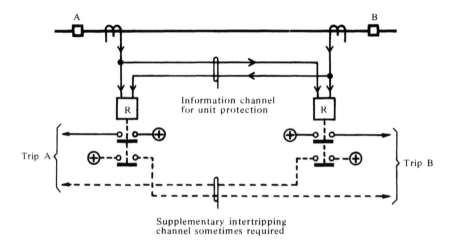

Fig. 10.2B   *Unit protection with large distance between c.t.s.*

10.3   Basic types of protection information channels

There are three main types of information channel used for protection at the present: (*a*) auxiliary conductors, for example, pilot wires: (*b*) the main conductors of the protected circuit or some other circuit running in parallel; (*c*) aerial transmission, for example, radio links.

The choice of a particular type of information channel depends upon many factors such as economics, availability of channels, the type and length of the protected line and other factors such as the possible uses (other than protection) for which the information channel may be required.

Some of the features and parameters of these types of information channels have been described in Chapter 7, together with associated terminal equipment, particularly where this is used in digital information systems. The protection systems covered in this Chapter are based mainly on analogue information and, although most of Chapter 7 is still relevant, some features related to analogue information are referred to in the following Sections.

10.3.1   Pilot wires

There is considerable variety in the types used, and the subdivisions of these different types cover such factors as core size, insulation level of the cores, whether the pilot is underground or overhead, and particularly whether the pilot is privately owned or rented. Some of the main details of a representative range of pilots are given in Table 7.2.2.1A of Chapter 7.

It will be noted that the intercore capacitance values are subject to some variation, the actual capacitance obtained for any given cable being dependent on the design of the cable and on the type of insulating material employed. Thus, telephone cables using plastic insulation (polythene or p.v.c.) generally have a higher value of intercore capacitance than paper-insulated cables, this being particularly so in the case of p.v.c. insulation.

10.3.2   Main conductors

The main conductors of the power system may be used in either of two ways. With a suitable h.f. coupling equipment, high-frequency signals may be injected and received using the main conductors as the actual communication link. There are, of course, many ways of superimposing information on such a basic h.f. carrier signal, and these will be described later. The frequency range available for this purpose is generally in the band of 70-500 kHz. The second way of using the main conductors is to disturb the system deliberately by placing a fault on it, for example by a fault-throwing switch, causing fault currents to flow and so causing operation of protective relays at remote points. This method is described in Chapter 17.

10.3.3   Radio links

Although not widely used for protection in the UK, their use has increased considerably in other countries over the last 10-15 years. They have particular advantages where large amounts of information are required for control and communication, and for some areas of difficult terrain. They depend on line-of-sight transmission, and beamed radio signals generally in the frequency band of 1000-3000 MHz, or even above. Again, various forms of information can be superimposed on the basic carrier for different types of protection.

10.4   Types of information used

There are different types of information which may be derived from the primary circuit and transmitted over the information channel. These range from complete information about both the phase and magnitude of the primary current to very simple information as, for example, when the action of some local relay is used to switch on or off a communication signal. These various forms of information and brief comments on their application are described below. They will be dealt with in more detail in the relevant Sections of this Chapter concerned with specific forms of protection. Those systems based on command signalling have been covered in detail in Chapters 7 and 9.

10.4.1   Complete information on magnitude and phase of primary current

The use of full phase and magnitude information automatically implies that the system is a true differential one. In practice, this is applicable mainly to pilot wires, and even then to special cases such as multi-ended circuits. It has also been used in carrier systems for special applications on important transmission lines abroad. The use of a differential system presents some difficulty in obtaining full information linearly over a wide range of current, but, in theory, the differential system has very few limitations.

Practical difficulties may impose the need to deviate from a true differential system. A particular example is when the secondary signals derived from the primary current are only directly proportional to the magnitude of the primary current over a limited range. This has the effect, at high currents, of losing the magnitude part of the information leaving phase-angle information only, as described in the next Section.

10.4.2   Phase-angle information only

Information as to the phase angle of primary current is relatively simple information, that is it is not dependent upon current magnitude, and therefore it is relatively easy to superimpose on and transmit over any type of channel. Pilot wires, carrier current over the primary conductors and, more recently, radio links, have all been used for this purpose. There are certain limitations in protection performance when phase-angle comparison is used as will be described in more detail later.

10.4.3   Simple two-state (off/on) information

This form of information can be used with all three types of channel. The function of the channel may be either to block tripping (secure stabilising during external faults) or to secure tripping under internal faults. Such tripping may be direct, in which case it can be classed as intertripping, or it may be permissive through some relay at the receiving end. The method in which the channel is used, and the type of channel which is used, will govern the various features and requirements of the protective system. The technique used and their application are described in Chapters 7 and 9.

10.5   Starting relays

These are necessary where the communication channel used for unit protection is not continuously employed. This may be the case, for example, in:

(a)   Pilot wire systems, where the pilots are normally supervised and such supervision is not compatible with the discriminating action of the communication channel. Alternatively, the channel may be used for general communication under normal system conditions.

(b)   Carrier systems where there are restrictions on continuous transmission of carrier power.

(c)   Any unit system where the principle used for discrimination is not valid at low-level fault conditions.

Starting relays are usually simple devices and are nondiscriminative because they are actuated by quantities existing only at the termination where they are located. They are sometimes called 'fault detecting' or 'sensing relays'. They do not do any tripping on their own account. However, they will usually decide fault settings, and their design and choice is important in this respect. They will also affect the overall tripping time of the protection, and this is also an important feature.

By way of illustrating their use consider a simple scheme, with a communication channel not normally in service (Fig. 10.5A). On the occurrence of a fault condition on the system, this channel must be changed over to protection use in order to permit the discriminating feature of the protection to decide whether the fault is internal or external, that is to trip or to stabilise. This switching is done by a starting relay at each end, which detects the fault condition.

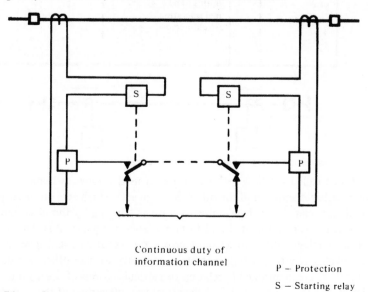

Continuous duty of
information channel

P – Protection

S – Starting relay

Fig. 10.5A   *Channel switching using starting relays*

If it is an internal fault, fed at one end only, the starting relay at that end will operate. As a result, the communication channel is connected at only this end and tripping will therefore occur at this end. This, however, is acceptable under these fault conditions.

If the fault is external, then stabilising is essential, and fault current is present (of approximately the same value) at both ends and we must then ensure that the starting relays operate at both ends.

This is easily done at high fault currents, in excess of the settings of the relays, but not easily at low level marginal fault conditions because of relay errors. The effect of this is illustrated in Fig. 10.5B.

It could give operation of relay A and not relay B, present the protection with an internal fault condition and so produce a maloperation. Such marginal fault conditions often arise on modern power systems.

The normal procedure for overcoming this difficulty is to fit two starting relays

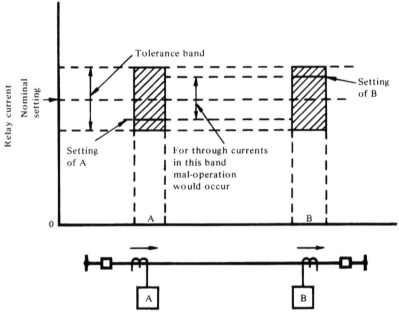

Fig. 10.5B   *Maloperation owing to starting relay tolerances*

at each end, one called the 'low set', which performs the above switching duty on the communication channel, and the other the 'high set', which acts on the tripping circuits of its associated end in a sense to *prevent* tripping by the discriminative protection when it is not operated and to *permit* such tripping if it is. The relative settings of such relays take into account errors and variations, as shown in Fig. 10.5C. In assigning relative settings it is often necessary to take other factors into account, but these are mentioned in relation to particular forms of protection.

The high-set relays, of course, define the effective fault setting of the protection, and the settings of such relays have to be chosen with some care.

As three-phase conditions are balanced, it is difficult to distinguish low-level three-phase faults from load conditions. The three-phase settings of starting relays must be chosen to exceed the maximum load conditions to be expected on the circuit. (There is an exception to this in the case of impulse starting of phase-

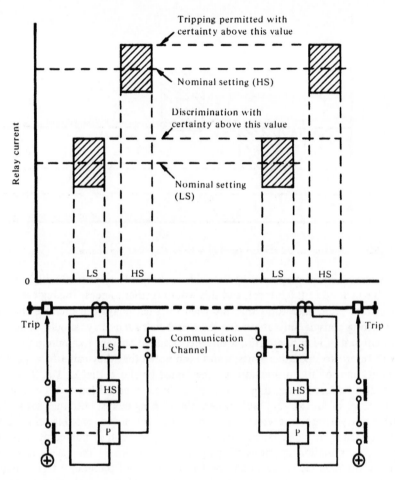

Fig. 10.5C *Application of high-set and low-set starting relays*

comparison carrier protection Section 10.10.9).

The unbalanced condition cannot be anything other than a fault condition, and the choice of setting is not therefore influenced by load conditions.

For a normal maximum load of 1·0 p.u., and possible overload of 1·1 p.u. as shown in Fig. 10.5D, then under no circumstances, must the discriminating channel be connected at this latter value. This means that the low-set relay must have a *reset* value greater than 1·1 p.u., say 1·25 p.u., to allow for relay errors. This immediately draws attention to the need for a high reset ratio (reset)/(pick-up) on the relay, if the overall settings of the protection are to be as low as possible. For an 80% reset, the pick-up of the low-set relay becomes about 1·5 p.u.

The pick-up of the high-set relay must not overlap that of the low-set, with the practical variations to be expected (and any other requirements such as capacity current), say a 20% margin, the pick-up of the high-set now becomes 1·2 x 1·5 = 1·8

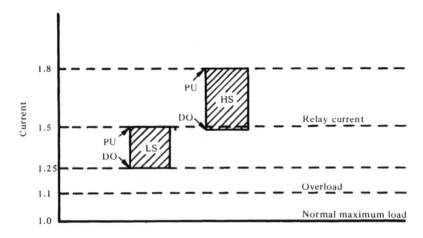

Fig. 10.5D   *Relationship between starting relay settings and load current*

and its reset 1·44. It may even be desirable to keep the reset of the high-set relay above the pick-up of the low-set, and this would further increase the high setting.

It can be seen, therefore, that the need for such relays can increase the three-phase setting considerably above the load, and that such relays should be accurate, consistent and of high reset ratio, if this increase is to be kept to a minimum.

With regard to operating times, it has been noted that they can affect the overall operating time of the protection, so they must be short, that is, 10-20 ms. They should also be fast in resetting, as it is possible for a condition to arise, following the clearance of an external fault, where the starting relays, both high-set and low-set have not yet reset at each end, and there is a load current existing on the circuit in the presence of capacitance current. If this condition is not compatible with correct discrimination spurious tripping may occur before the relays have reset. Certainly the high-set relays should be fast to reset and preferably faster than the low-set. In some cases the reset times are deliberately controlled to be so. The problem of fast resetting remains, however, and is greater the shorter the operating time of the discriminating link becomes.

It can be seen that the choice, design and application of starting relays need care, and present a different picture from the elementary idea of a simple electro-magnetic relay with a nominal setting. Structural aspects such as high insulation, elaborate contact assemblies, and special reliability requirements for some contacts expand the problem further.

10.6 Conversion of polyphase primary quantities to a single-phase secondary quantity

10.6.1 General philosophy

With few exceptions, it is general practice to use a single-phase quantity as the information to be conveyed over the communication channel; this reduces its complexity and cost. This technique is generally justified because the primary fault currents bear definite relationships for the various types of faults, and because the same summation technique is used in an identical manner at the several ends of the protected feeder. Some reference to these summation techniques will have been given in previous Chapters. They are summarised here with particular reference to unit protection of feeders, and more details of individual circuits will be given when particular forms of protection are described.

10.6.2   Interconnections of current transformers

The secondary windings of current transformers may be directly interconnected to provide single-phase output currents. For example, the parallel connection of current transformers will provide a single current output which is related to the zero-sequence primary current. This is, in effect, a simple form of zero-sequence output circuit (Fig. 10.6.2A). By providing different numbers of secondary turns

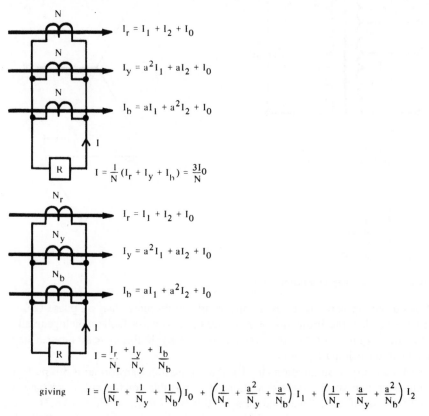

$$I_r = I_1 + I_2 + I_0$$

$$I_y = a^2 I_1 + a I_2 + I_0$$

$$I_b = a I_1 + a^2 I_2 + I_0$$

$$I = \frac{1}{N}(I_r + I_y + I_b) = \frac{3 I_0}{N}$$

$$I_r = I_1 + I_2 + I_0$$

$$I_y = a^2 I_1 + a I_2 + I_0$$

$$I_b = a I_1 + a^2 I_2 + I_0$$

$$I = \frac{I_r}{N_r} + \frac{I_y}{N_y} + \frac{I_b}{N_b}$$

giving
$$I = \left(\frac{1}{N_r} + \frac{1}{N_y} + \frac{1}{N_b}\right) I_0 + \left(\frac{1}{N_r} + \frac{a^2}{N_y} + \frac{a}{N_b}\right) I_1 + \left(\frac{1}{N_r} + \frac{a}{N_y} + \frac{a^2}{N_b}\right) I_2$$

Fig. 10.6.2A   *Parallel connection of current transformers*

on each of the phases an equivalent output is obtained which will depend upon the type of fault and the relative ratios of the secondary turns. These methods of current summation are not used frequently for unit feeder protection as the secondary current levels of the current transformers are not really suitable for comparison over the information channel. Once it is necessary to transform the current level, a current summation device is possible, as described in the next two Sections.

10.6.3   Summation transformers

These devices were frequently used for feeder protection, particularly those involving pilot wires. The general arrangements are shown in Fig. 10.6.3A, from which it can be seen that it is possible to obtain a single phase output for all the common types of faults. It is also possible to control independently the outputs for earth faults and phase faults.

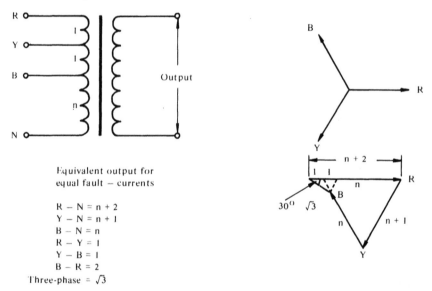

Equivalent output for
equal fault – currents

$R - N = n + 2$
$Y - N = n + 1$
$B - N = n$
$R - Y = 1$
$Y - B = 1$
$B - R = 2$
Three-phase $= \sqrt{3}$

Fig. 10.6.3A   *Summation transformer*

The output on earth faults is usually considerably more than on phase faults, which means that the protection is more sensitive to earth faults which generally satisfies the requirements on the system, particularly if this is resistance-earthed with a limited earth-fault current.

As with all current summation circuits, the summation transformer is not perfect and under certain special fault conditions it is possible for the output to be too small to be significant. Two examples of this are shown in Figs. 10.6.3B and 10.6.3C. In general, however, it is possible to choose tappings which make such

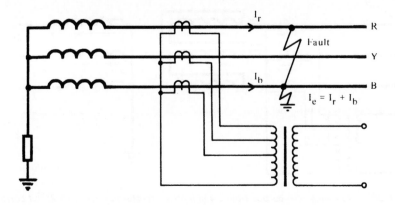

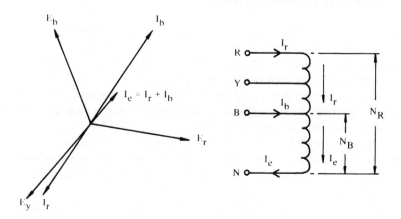

Total ampere − turns $AT = (N_R − N_B) I_r + N_B I_e$

$$= (N_R − N_B) (I_e − I_b) + N_B I_e$$

from which $AT = N_R I_e − (N_R − N_B) I_b$

Assuming I_b and I_e to be in phase, the condition
for zero ampere − turns is: −

$$\frac{I_b}{I_e} = \frac{N_R}{N_R − N_B}$$

Fig. 10.6.3B *Showing zero output from a summation transformer under a two-phase-to-earth fault condition*

cases very unlikely, especially on multiple solidly earthed systems. There is always the occasional combination of system conditions which might give rise to difficulty, and these must receive special analysis.

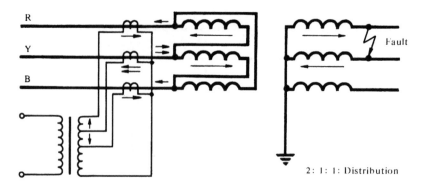

Fig. 10.6.3C   *Showing zero output from a summation transformer owing to 2:1:1 current distribution in the power transformer*

10.6.4   Phase-sequence current networks

This class of device is much more complex than a summation transformer and is consequently more complicated to design and expensive to produce. It uses inter-connection of the various current transformer secondaries and the inclusion of complex impedances, the combination having the ability of filtering out one or more of the sequence components of the fault current and so give an output which is related to particular chosen sequence components. A simple example is shown in Fig. 10.6.4A, but there are many variations.

Most sequence networks have been based on passive components, e.g. trans-formers, reactors, capacitors and resistors, but increasing use is being made of active components to amplify and/or combine the electrical quantities and some-times to produce the required phase shifts.

As sequence networks use phase-shift circuits, they are really steady-state devices. When used for instantaneous types of protection therefore, their response to transient conditions and non-power frequency conditions must be considered in the design. As will be shown later, it is usually necessary to build in special frequency filters to avoid spurious outputs.

As in the case of summation transformers, there is the possibility that certain system conditions will result in a reduced or negligible output. The characteristics of such networks, therefore, have to be chosen very carefully in relation to the system conditions in order to minimise the danger of such failures.

The expense and complication of sequence networks generally prohibits their use for pilot wire protection, and they are usually reserved for more elaborate forms of protection using carrier techniques.

10.7   Elementary theory of longitudinal differential protection

Before proceeding to a detailed study of the particular forms of unit protection for

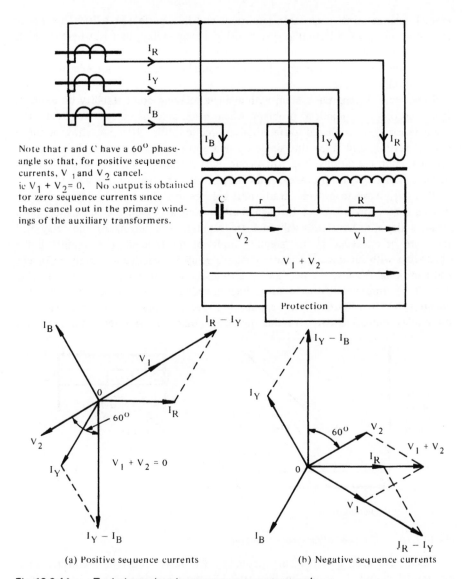

Note that r and C have a 60° phase-angle so that, for positive sequence currents, V_1 and V_2 cancel. ie $V_1 + V_2 = 0$. No output is obtained for zero sequence currents since these cancel out in the primary windings of the auxiliary transformers.

(a) Positive sequence currents

(b) Negative sequence currents

Fig. 10.6.4A   *Typical negative-phase sequence current network*

feeders this section considers some elementary principles which are necessary for a better understanding of the problems involved and the techniques used.

10.7.1   Longitudinal differential protection with biased relays

With perfect c.t.s and symmetry of connection the basic differential system gives perfect balance on through fault conditions and thus perfect stability. This is only true of an unbiased differential system using a single relay, that is, where the ends of the protected zone are fairly close. It will be seen that this perfect theoretical balance does not exist on longitudinal differential protection on a feeder where the ends are well separated.

Even for a differential system with a single relaying point stability on external faults becomes a practical problem when instantaneous relays are used. The characteristics of unbiased differential protection, Fig. 10.7.1A, show a single current setting, and since this single current setting must cater for the minimum internal fault conditions that can occur on a system, it will be well below the maximum through current on external faults for which stability is required.

It is fairly easy to provide steady-state balance in a differential system but there will be some upper level of through fault current at which the unbalance, both steady-state and transient, will increase rapidly. The setting of the relay will therefore be exceeded at the 'stability limit' of the scheme. The stability limits obtainable with an unbiased low-impedance relay in practice are rarely sufficient and additional features are necessary, particularly with instantaneous schemes. One way of securing stability is to use a high-impedance relay, as discussed in other chapters, but this solution is not applicable to feeder protection. Another way is to use a relay which has two actuating quantities, that is a biased relay, various forms

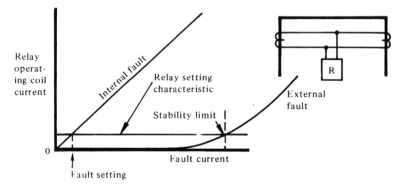

Fig. 10.7.1A   *Unbiased-differential protection*

of which have been described in Chapter 6. The relay would be connected in a differential circuit, as shown in Fig. 10.7.1B, the operating winding being in the usual residual connection and the other winding, the restraining winding, connected in the circulating path between the c.t.s. The current required to operate the relay will be increased by some controlled amount with increasing circulating current and so extend the stability limit for through faults. The setting of the relay will be increased on internal faults, but this will be relatively small provided the bias characteristics are chosen correctly.

The presence of a through current at the same time as an internal fault current will, of course, increase the setting further, as shown in Fig. 10.7.1C. Such a con-

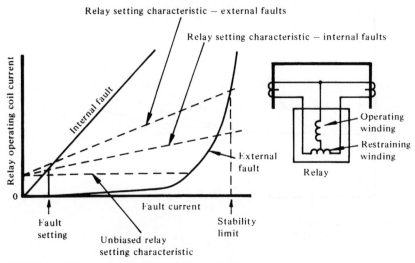

Fig. 10.7.1B *Simple biased-differential protection*

dition can occur for a low level fault which did not collapse the system and reduce the load. By delaying the growth of the bias characteristic, as shown in Fig. 10.7.1D, it is possible to ease the problem of increased marginal settings at the same time as being able to provide the necessary stability. Here, it must be emphasised that the main problem with biased relays is to choose a bias characteristic which is sufficiently in excess of the out-of-balance current on through faults to ensure stability, and yet sufficiently below the fault current available on internal faults to ensure fast positive action. The ease of doing this depends on the separation between the curves of current in the operating winding for internal and external fault conditions.

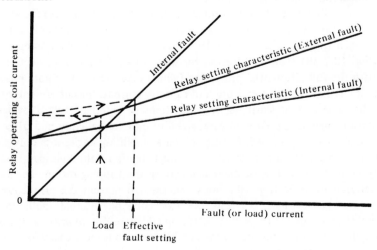

Fig. 10.7.1C *Relay with linear bias characteristic*

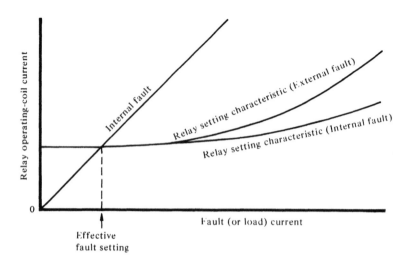

Fig. 10.7.1D   *Relay with nonlinear bias characteristic*

10.7.2   Phase-comparison principles

As previously noted, the principle of phase comparison is particularly applicable to the protection of feeders with either pilot wires or carrier. There are some elementary principles which are worth appreciating, although these will be amplified later. Considering a simple system where it is possible to compare the phases of the currents at the two ends of the protected circuit, as shown in Fig. 10.7.2A, it is seen that irrespective of the magnitudes of these currents the output signal of a phase comparator would vary as shown.

It is seen, therefore, that according to the setting of this comparator, the characteristic can be set to have a particular angle between the currents at the two ends at which tripping takes place and a complementary angle where stability takes place (Fig. 10.7.2B). The justification for using this principle of protection compared with the full differential protection (which takes account of both phase and magnitude) is that, with the exception of capacity current, the current entering the line is the same as that leaving it for conditions of external fault. Provided the characteristic angle is chosen with care, stability on external faults and tripping on internal faults can be secured. Having chosen a stabilising angle, it is only safe to permit comparison at current levels on through faults or through load when the capacity current will not cause encroachment on the tripping region. The starting relays referred to in Section 10.5 are often used to prevent this erroneous comparison. The phase angle between the two ends on internal faults may be considerable and difficult to predict. It is normal, therefore, to arrive at the phase-angle requirements from the stabilising condition which is more easily defined, and to adopt as large a tripping angle as possible.

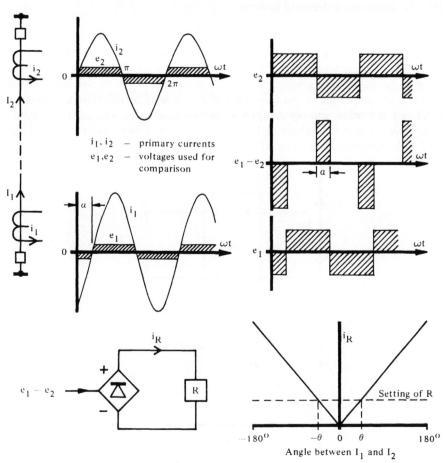

Fig. 10.7.2A   *Basic conept of phase comparison protection*

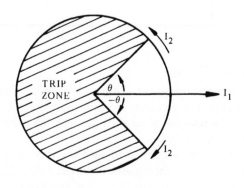

$^+_-\theta$ — stability angle

Fig. 10.7.2B   *Characteristic of phase comparison protection*

10.7.3   Nonlinear differential systems

A system may begin as a differential system taking account of both phase and magnitude. This means that the secondary quantities which are derived from the primary quantities vary proportionally at low current levels. As the current level increases the amplitude of the quantities may be limited by magnetic saturation or non-linear resistance, and the comparison deteriorates to the only other remaining information, which is of phase-angle difference between these limited signals (Fig. 10.7.3A). This principle is frequently used.

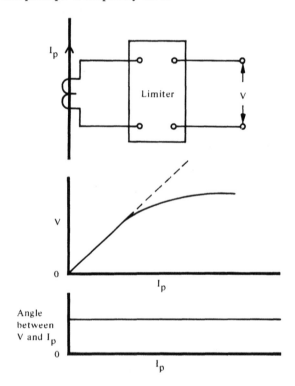

Fig. 10.7.3A   *Magnitude limiting*

In a pure phase-comparison system, the amplitude of the currents is ignored and, hence, if there is a capacitance current fed into the protected circuit at a low level of through-current, the resulting phase-shift between the currents at the two ends can vary, as shown in Fig. 10.7.3B.

If a fixed tripping angle is assigned to such a system of protection, there will obviously be a lower limit of through-fault current below which the phase angle between the currents at the two ends of the line will exceed the stabilising angle, thus tending to produce unwanted operation of the protection.

Amplitude-sensitive relays such as starting relays, must therefore be provided at each end to prevent such unwanted operation. The problem is rather more complicated than this simple treatment suggests and is treated more fully in the Section on carrier protection.

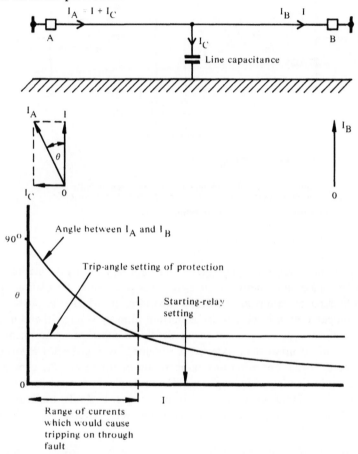

Fig. 10.7.3B   *Incorrect comparison under external fault conditions owing to line capacitance*

In a hybrid scheme, namely one which operates as a simple differential system at low current levels and as a phase-comparison system at higher current levels, we can arrange for limiting to take place at a level which automatically safeguards against maloperation at low levels. The capacitance current would then appear as an out-of-balance in the unlimited working range of the protection, and we would ensure that the setting of the differential relays was sufficiently in excess of this to prevent maloperation. Such a system, therefore, can be considered as containing its own amplitude-sensitive feature which, in a pure phase-comparison system, would have to be provided by additional starting relays.

True phase comparison is really permissible only for a two-ended protected zone. Where the protected zone has three or more ends, such as tapped or tee'd circuits, the principle fails or is very dependent upon particular system conditions

(Fig. 10.7.3C). It is seldom applied to such circuits.

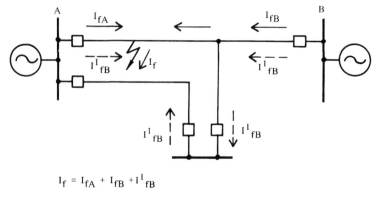

$$I_f = I_{fA} + I_{fB} + I^1{}_{fB}$$

Fault current leaves protected zone on internal fault -
amount depends on relative primary impedances to fault

Fig. 10.7.3C   *Incorrect comparison on three-ended feeder*

10.7.4   Directional comparison systems

A pure phase-comparison principle requires the transmission of information of the actual relative phase angle between the two primary currents. A variation on this theme is obtained by using directional comparison, in which the phase of the current is compared with respect to its associated primary voltage, the same thing happening at both ends. Directional relays do this comparison. Assessment as to whether the fault is internal or external is obtained by transferring information between ends as to the operation or non-operation of the relays, the communication link acting in a simple switching sense. The principle is not so elegant as phase-comparison and chronologically preceded it. It is subject to similar limitations with respect to capacitance current and is similarly not very suitable for multi-ended circuits.

10.7.5   Current sources and voltage sources

It is important in protection to recognise the difference between a 'current source' and a 'voltage source' as both are met with in practice. A current source is shown in Fig. 10.7.5A and has the characteristic that the current through a connected load circuit is relatively unaffected by the impedance of this load circuit. Looking into the terminals from the load it can be imagined as consisting of a very high voltage driving into a very high impedance. A current-transformer is a practical case and the two forms of equivalent circuit, namely the current-source equivalent and the voltage-source equivalent, are shown in Fig. 10.7.5B.

    If two currents are to be compared, they may be fed into a single point sometimes called a 'junction' and the summated current leaving this junction monitored

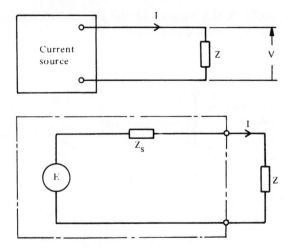

V = ZI. For constant current, V varies with Z

E is large and $Z_s \gg Z$

So that $I = \dfrac{E}{Z_s + Z} \simeq \dfrac{E}{Z_s}$

Fig. 10.7.5A   *Current source and its equivalent circuit*

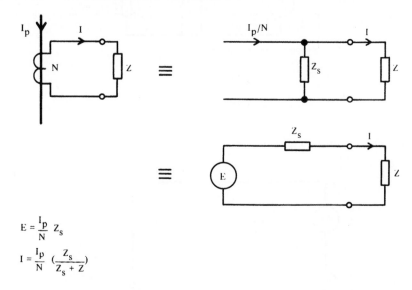

$E = \dfrac{I_p}{N} Z_s$

$I = \dfrac{I_p}{N} \left(\dfrac{Z_s}{Z_s + Z}\right)$

Accuracy depends on $Z_s \gg Z$, whence

$I = \dfrac{E}{Z_s + Z} = \dfrac{1}{N}I_p Z_s . \dfrac{1}{Z_s + Z} \simeq \dfrac{1}{N}I_p$

Fig. 10.7.5B   *Current transformer equivalent circuit*

by a relay (Fig. 10.7.5C). It is easy to see the practical equivalent to this in a differential connection of current transformers.

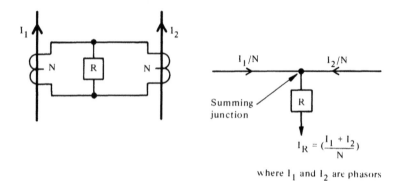

Summing junction

$$I_R = (\frac{I_1 + I_2}{N})$$

where I_1 and I_2 are phasors

Fig. 10.7.5C   *Comparison of two currents*

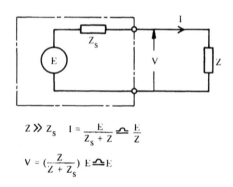

$$Z \gg Z_s \quad I = \frac{E}{Z_s + Z} \simeq \frac{E}{Z}$$

$$V = (\frac{Z}{Z + Z_s}) E \simeq E$$

Fig. 10.7.5D   *Voltage source equivalent circuit*

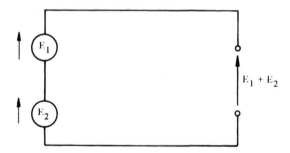

$E_1 + E_2$

Fig. 10.7.5E   *Summation of voltages*

A 'voltage source' is defined as one in which, if a load impedance is connected, then the current flowing in this load impedance is directly proportional to the impedance value. If two voltages are to be compared they are connected in series and either the total voltage or the current flowing in a connected impedance (Fig. 10.7.5E) is measured.

Both forms of source have errors in practice, and it can be seen that this error (or a measure of the goodness of the source) is directly dependent on the relative magnitude of the source impedance and the load impedance.

It is rather arbitrary as to whether the secondary quantity related to a primary current is derived as a voltage source or a current source. The deriving of a secondary current is more familiar, but in feeder protection it is sometimes more convenient to derive a secondary voltage source.

A current source is relatively easily converted into a voltage source by feeding the current into an impedance which is low compared with the impedance to be connected to it (Fig. 10.7.5F). This is not a perfect voltage source but sufficient for practical purposes.

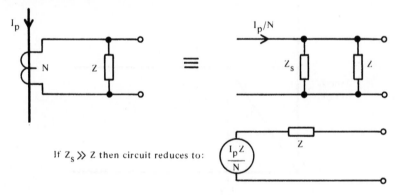

If $Z_s \gg Z$ then circuit reduces to:

Fig. 10.7.5F   *Converting a current source to a voltage source*

10.7.6   Nonlinearity and limiting

As has already been mentioned, this is an important feature of many forms of feeder protection. To appreciate the mode of operation of some systems it is necessary to understand the two principal forms of nonlinearity used.

(*a*)   *Limiting due to saturation of iron circuits:* Some of the features of magnetic saturation have been dealt with in connection with current transformers in Chapter 4, but they are repeated here for the sake of clarity. Taking a simple iron circuit, shown in Fig. 10.7.6A, which is excited by an a.c. current, then there are a number of quantities which can be plotted to a base of exciting ampere turns, namely flux density, average voltage, peak voltage and r.m.s. voltage. If a simplified BH characteristic of the material is assumed then at some specific value of ampere-turns the material will be incapable of developing any appreciable increase in flux density, that is, the iron becomes saturated. This means that the average excursion of flux

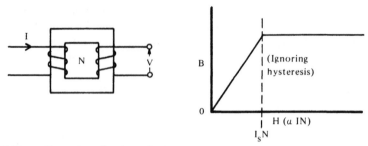

Fig. 10.7.6A   *Saturation of an iron circuit*

in the winding becomes constant and the average value of voltage, irrespective of its waveform, will be limited, as shown in Fig. 10.7.6B. It will be clear from Fig. 10.7.6C that at high currents the whole of this flux change will occur around the region of current-zero and that the duration of voltage output will be short, but that the amplitude may be high. Ideally, the area of the voltage waveform will be

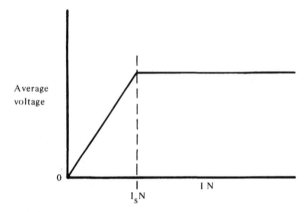

Fig. 10.7.6B   *Limiting of average output voltage*

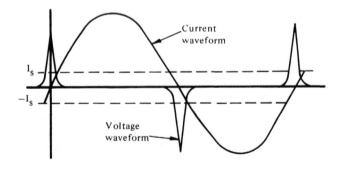

Fig. 10.7.6C   *Voltage output with severe saturation*

constant, which is the same as saying that the average value of voltage is constant. It is clear from this that the peak value of voltage will continue to rise, and in an ideal system would be proportional to the slope of the current input which, of course, would be proportional to the peak of the input current, assuming it is a sine wave. In practice, hysteresis and eddy-current losses together with the effects of connected impedance will limit the rate of increase of peak voltage (Fig. 10.7.6D).

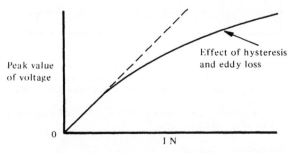

Fig. 10.7.6D   *Variation of peak voltage with saturation*

In practice, this peak voltage rise is some fractional power of the input current (0·2-0·5). Nevertheless, an iron circuit, which may have a limited average voltage of say 50V, may well develop peak voltages of the order of 1 or 2 kV at high currents. The r.m.s. value of voltage, which would often be measured, also would be limited but not quite so effectively, as it would rise gradually after saturation. This is because the r.m.s. value of a quantity is dependent upon its waveshape even though its average value may be constant. There are a number of important aspects of iron saturation to take note of:

 (i)   Peak voltage values are not limited.
 (ii) The waveform is considerably distorted and peaky, containing many harmonics.
 (iii) The peaky output voltage is located around current zero and can therefore be considered as defining the phase of the input current by a short duration pulse.
 (iv) The value of voltage that we will measure under such conditions will be dependent upon the type of instrument that is used.

(*b*)   *Nonlinear resistors:* If a current transformer is loaded by one of the nonlinear resistance materials, for example, Metrosil, then limiting takes place with an entirely different action. The d.c. characteristic of the Metrosil material shows that the voltage and current are related by a power law, as in Fig. 10.7.6E. For all practical purposes, the d.c. curve applies for instantaneous values of an a.c. condition. If a high level sine wave of current is applied to such a combination as shown in Fig. 10.7.6F, the voltage developed across the non-linear resistor will be a flattened waveform, and the peak value will be limited and will only rise slightly as will also the average and r.m.s. values. Some important observations about such a system of

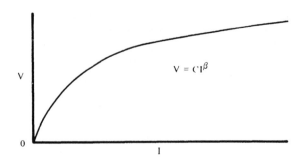

$$V = C I^\beta$$

C is a constant expressed in volts dependent on the dimensions and material of the metrosil unit and is the voltage developed across the metrosil at a current of 1 amp.

β is a constant, ranging from 0.25 to 0.3, and is dependent on the material only.

Fig. 10.7.6E   *Characteristic of metrosil*

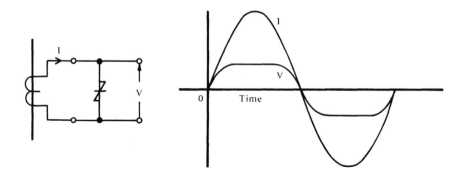

Fig. 10.7.6F   *Voltage waveform with limiting by nonlinear resistance*

limiting may be made.
(i)   Average values are limited.
(ii) Peak values are limited.
(iii) The duration of the voltage wave is the same as the input current.
(iv) The output voltage defines the phase relationship of the primary current by flattened voltage wave of the same duration.

Generally it is found that systems of the older type use saturation of iron and pilots with an adequate insulation level to withstand the high peak voltages on the pilots, whereas more modern systems developed for use with telephone type pilots will favour the use of nonlinear resistors. The use of either method, however, has the effect of removing the amplitude relationship to the input current and retaining only the phase relationship.

10.8   Pilot-wire protection

10.8.1   Basic principles

The application of the differential principle to feeder protection merits consideration so that the characteristics of various systems, and the factors influencing their
design, may be appreciated. Starting with the basic differential system shown in
Fig. 10.8.1A, the impedance of the connecting leads is low compared with the
current transformer winding resistance, and the VA burden is correspondingly low.
As the interconnections are short the total intercore capacitance will also be low.
With such a simple arrangement it is possible to provide one relaying point which is
located at the equipotential point of the interconnecting leads and which can
control both circuit breakers.

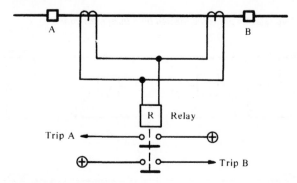

Fig. 10.8.1A   *Basic differential system using one relay*

With the considerable separation between the circuit breakers controlling the
two ends of a feeder, there is a need to have at each end a relaying point controlling
its associated circuit breaker, these relaying points being connected differentially.
This need for two relaying points means that the conventional midpoint relay is no
longer applicable and some other connection is necessary to retain the quality of
discrimination between internal and external faults. This connection must be
symmetrical, that is, the same at both ends.

Keeping the basic principle of differential current balance, a simple relay connection can be obtained by using two relays and a third pilot core, as shown in
Fig. 10.8.1B.

The relays are in series with this third core, and both are energised during internal
faults fed from either end or both ends. For no pilot capacity, the relay connection
is across equipotential points on the pilot loop on external faults, and so is
unenergised.

The arrangement described is a true differential current system, using a third
pilot core to overcome the basic problem of the relaying connection. This principle
was used in the early applications of unit feeder protection.

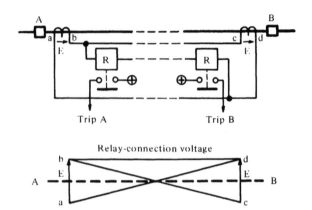

Fig. 10.8.1B   *Two relaying points connected by a third pilot core*

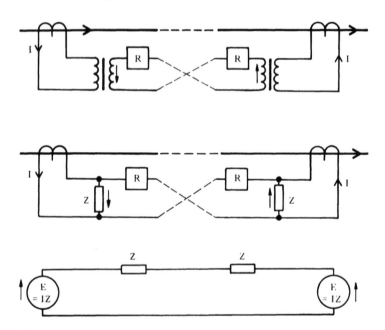

Fig. 10.8.1C   *Principle of voltage balance*

It was realised early in the history of feeder protection that an alternative solution to the problem of a suitable relaying connection was the use of a 'voltage balance' arrangement. In such an arrangement the secondary currents are replaced by, or converted to, an equivalent secondary voltage source of fairly low impedance, and these are then compared round a pilot loop of two cores, as shown in Fig. 10.8.1C (see previous Section on current source and voltage source.)

Any comparatively low-impedance load across the current transformer would serve the purpose of converting a secondary current into a secondary voltage and of providing the necessary low source impedance. In early systems, an air-gap introduced into the current transformer core was used for this purpose. In later systems, air gaps in auxiliary summation transformers or loading impedances produced the same effect. It should be noted that the voltage balance system is still fundamentally a differential system and it is only the secondary mechanism of balance which has changed.

10.8.2 Practical relay circuits

In practice, the presence of pilot capacitance and current transformer inaccuracies require some modification to the basic circuits previously described. In simple differential protection where the two ends are close together, pilot capacitance is negligible and inaccuracies are overcome either by using biased relays or high impedance relays (see Section 10.8.1). High-impedance relays are not generally applicable to feeder protection; biased relays are and appear in many practical systems. The application of biased relays can be split into two broad classes:
(*a*)   Those having true current source secondary energisation such as in Fig. 10.8.1B
(*b*)   Those having secondary voltage source energisation such as in Fig. 10.8.1C.
    In the first class, bias is derived from the current circulating round the pilots during through faults. It is therefore equivalent to a simple biased differential system.

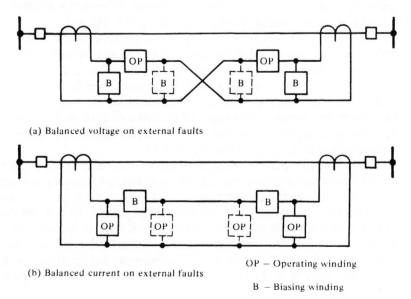

(a) Balanced voltage on external faults

(b) Balanced current on external faults

OP – Operating winding

B – Biasing winding

Fig. 10.8.2A   *Alternative arrangements for biased-differential feeder protection*

In the second class, we can take a different approach to understanding the function of a biased relay. The connection of such a biased relay is entirely local and there are thus only two quantities which can be used for operation or bias, namely the current in the pilots and the voltage across them. The relay receives information about the conditions at the far end due to the effect that the far end secondary voltage has on the voltage and current at the relaying point.

With this general approach to the second class, there are two possibilities, as shown in Fig. 10.8.2A. In one case, (*a*), the pilot voltages oppose around the pilot loop during external faults. In this case the operating quantity on the relay is pilot current and the biasing quantity is pilot voltage. In the other case, (*b*), the pilot voltages are arranged to be additive during external faults thus reverting to a circulating current principle but with voltage source. The pilot current is then the biasing quantity, and the pilot voltage the operating quantity.

The various basic principles described have advantages and disadvantages when applied to practical systems using practical pilot circuits. These have led to considerable variety of detail in proprietary forms of protection and some change in fashion as the needs of power systems have changed, or the introduction of new components or techniques have favoured one arrangement or another.

10.8.3   Summation circuits

These have previously been described and it will be sufficient to say that the device generally used is a simple summation transformer or its equivalent.

In addition to the function of current summation to provide a single-phase quantity, the summation transformer may have other uses as follows:

(*a*)   *Ratio changing:* The high values of pilot loop impedance, for example, up to 2000 Ω or more, make the direct connection of pilots and current-transformer generally unsuitable. It is necessary therefore to change the current level between the pilot loop and the current transformer secondaries, and this is usually done by stepping down the current level on the pilot side of the summation transformer. Turns ratios of the order of 1:10 or more may be used thus reducing the burden that the pilot circuit imposes on the current transformers to an acceptable level (Fig. 10.8.3A).

(*b*)   *Isolation of pilots:* The two windings of the summation transformer can be used to isolate the pilots from the current-transformers. Specifications usually require current transformer secondaries to be earthed. Pilots are unearthed. The isolation between the pilots and the current transformers will therefore prevent a spurious earth connection between ends. Such a connection would be very undesirable. The provision of high insulation level on the summation transformer can also isolate the current transformer circuits from possible overvoltages induced in the pilot, or earth potential differences between the two ends.

(*c*)   *Conversion from current to voltage:* It has previously been noted that one

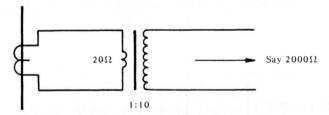

Fig. 10.8.3A   *Use of summation transformer to give change in impedance*

method of doing this is to introduce an air gap into the magnetic circuit of
the summation transformers, the latter being fed from solid core current
transformers. Impedance loading of the secondary winding of the summation
transfomer can be used as an alternative method (Fig. 10.8.3B).

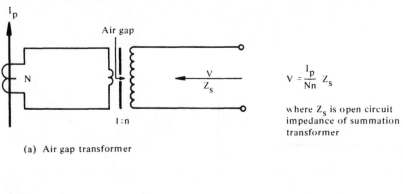

where Z_s is open circuit
impedance of summation
transformer

$$V = \frac{I_p}{Nn} Z_s$$

(a) Air gap transformer

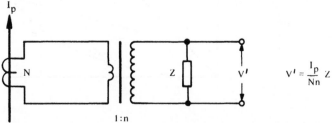

$$V' = \frac{I_p}{Nn} Z$$

(b) Low value secondary impedance

Fig. 10.8.3B   *Conversion from current source to voltage source*

10.8.4   Basic discrimination factor

The discriminating quality of a differential system can be assessed in terms of a
factor defined by the ratio of the degree of correct energisation of the relay under
internal fault conditions to that which occurs (and is unwanted) under external
fault conditions at the same primary current. In a perfect system, this factor would

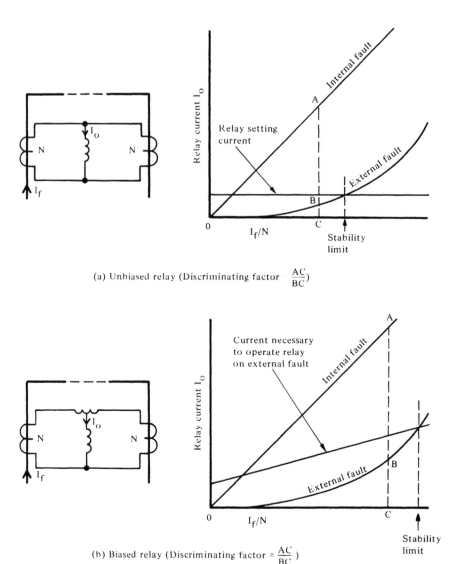

(a) Unbiased relay (Discriminating factor $\frac{AC}{BC}$)

(b) Biased relay (Discriminating factor = $\frac{AC}{BC}$)

Fig. 10.8.4A   *Discriminating factor*

be infinity and this is theoretically approached in a simple differential system. In practice, differences in the characteristics and loading of the current transformers reduce this factor. This is shown in Fig. 10.8.4A which gives the characteristics for (*a*) an unbiased relay, and (*b*) a biased relay. In the case of feeder protection, even if these practical limitations are ignored, the maximum discriminating factor possible is governed by the characteristics of the pilot.

Part 1, Section 10.7.1 of the chapter, showed the characteristics of a differential system by plotting the operating current in the relay for both external and internal fault conditions, together with the setting or bias characteristic of the relay, against a base of fault current. If a discriminating factor is assigned, then it can be seen that the ease with which a relay characteristic can be fitted to satisfy both the internal and external fault conditions (with an adequate margin for both conditions) is directly dependent on the value of the discriminating factor. With simple differential systems the choice of relay characteristic is relatively easy, since the theoretical discriminating factor is inherently high. Practical inaccuracies increase the problem but a satisfactory characteristic can usually be chosen.

In differential feeder protection, the theoretical discriminating factor is limited by the pilots and in extreme cases can create a problem of choosing a satisfactory relay characteristic. It is useful to consider the relationship between pilot capacitance and discriminating factor for the two basic types of circuits as follows:

(*a*) *Current-balance system:* With a simple current-balance system, as shown in Fig. 10.8.4B, the capacitance of the pilots does not introduce unbalance on external faults but will slightly reduce the current available under internal faults. The basic discriminating factor is still infinity, and the pilots do not play a great part in this case especially as the pilot lengths are short.

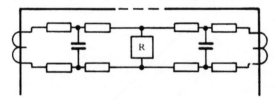

Fig. 10.8.4B   *Effect of pilot capacitance with two-core pilot system*

(*b*) *Voltage source systems:* Where secondary voltage sources are used to energise the pilots, the effect of pilots on discriminating factors may be examined in general terms. Two schemes are possible depending on whether the voltages are in opposition or in addition round the pilot loop under external fault conditions. In both cases the relationship between the pilot voltage and pilot current is the criterion which governs discrimination between internal and external fault conditions. With the basic circuit, as shown in Fig. 10.8.4C, three simple conditions of energisation can be used, namely:

(i)   from one end only;

(ii)   from both ends, the voltages being in opposition round the pilot loop; and

(iii)   from both ends, the voltages being in addition round the pilot loop.

The first condition must correspond to an internal fault fed from one end only. The second and third conditions will correspond either to an internal fault fed from both ends or to an external fault, according to the polarity of the interconnection. In all cases, the condition governing the operation or non-operation of the relay will be the ratio of pilot voltage to pilot current or its reciprocal.

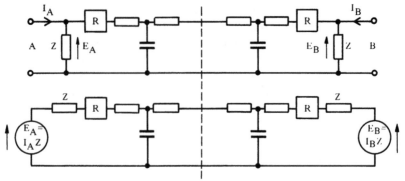

Main conditions:
(a) $E_B = 0$ .... Internal fault fed from end 'A'.
(b) $E_A = E_B$ .. Voltage balance: pilots can be considered
                   open-circuited at mid-point.
(c) $E_A = -E_B$ .. Circulating current: pilots can be
                   considered short-circuited at mid-point.

Fig. 10.8.4C   *Effect of pilot capacitance with voltage balance system*

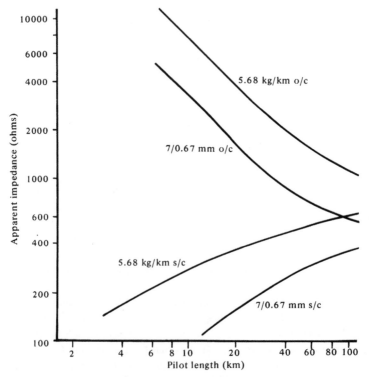

Fig. 10.8.4D   *Open-circuit and short-circuit impedance characteristics of typical pilot circuits*

This ratio is the effective pilot impedance or admittance seen by the relay at the pilot terminals under the various fault conditions. This is strictly true only if we ignore the unbiased setting of the relay, and this is permissible in practice on most relays, particularly at high fault currents.

This useful generalisation enables the discriminating factor to be expressed in terms of the impedance characteristics of the pilots, making allowance for the impedance with which the pilots are terminated. Typical open-circuit and short-circuit characteristics for two types of pilots are shown in Fig. 10.8.4D.

From Fig. 10.8.4C it will be appreciated that the open-circuit and short-circuit impedances are important because:

(a)   The short-circuited impedance of the whole pilot length represents the impedance seen by the relay on a single-end fed internal fault (assuming zero termination impedance).

(b)   The open-circuit impedance of half the pilot length is the impedance seen by the relay under external faults, on a voltage balance system.

(c)   The short-circuit impedance of half the pilot length is the impedance seen by the relay under external faults on a current-balance scheme using voltage sources on the pilots.

The separation between the two curves (open-circuit and short-circuit) of a pilot is therefore a measure of discriminating ability. This is only true if the relay

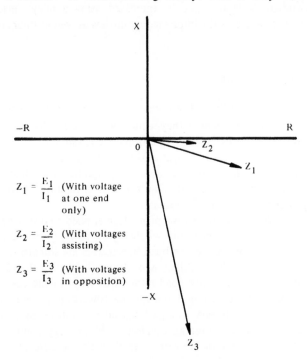

Fig. 10.8.4E   *Complex impedance diagram showing effective input impedance of a pilot circuit for various conditions of pilot energisation*

concerned is responding only to the magnitude and not phase-angles of the quantities.

If the pilots are long enough the impedance seen at the terminals will approach the characteristic impedance, and the relay at one end will not be able to detect either short-circuit or open-circuit condition at the other. In other words, the system may lose all its discriminating ability unless some additional feature is introduced.

Various discriminating factors are possible with practical lengths of pilot, and the case of the 5·68kg/km conductor pilot shows the order of difficulty encountered.

Fig. 10.8.4E shows approximately to scale the complex impedances which occur at the pilot terminals under the three basic conditions. The lowest discriminating factor for each type of circuit would be obtained by taking the single-end internal-fault impedance Z_1 and comparing it with the corresponding external-fault impedance. For the system where voltage opposition corresponds to an external fault, the discriminating factor would be Z_3/Z_1. For the case where voltage addition (current balance) corresponds to external faults, the discriminating factor would be Z_1/Z_2. Both these quantities may have the same approximate scalar value but in the former case there is a bigger phase angle difference between the two quantities. The fact that the two quantities may have this phase difference is important as the effective discriminating factor may be increased using a relay device which is sensitive to the phase angle of the impressed quantities as well as their scalar ratio.

10.8.5   Typical pilot circuits

(a)   *7/0·67 underground pilots:* These were among the earliest to be applied and grew up with cable systems where they could be economically laid at the same time as the main cables, thus avoiding additional trenching costs. The core section was 7/0·67, similar to that used in local auxiliary wiring. Insulation was about the same level, that is 2 kV test. The cores used for protection are usually wormed to reduce inductive interference. To suit some types of protection, separate screening of individual protective cores may be required.

(b)   *Overhead pilots:* For a differential protection applied to overhead lines where the pilots would have had to bear the whole laying costs, overhead pilots strung either on separate poles or on the overhead line towers are sometimes used. In the latter case they are frequently combined with the earth wire for the circuit to form a composite cable known as an overhead pilot earth wire. In the case of overhead pilots, and especially those strung on the same towers or poles as the primary circuit, particular care is necessary as high longitudinal voltages may be induced in the pilots because of mutual inductance between the primary conductors and the pilot. This condition is severe in the case of earth faults because of the high zero-sequence coupling. A wormed arrangement for the protective cores is necessary to avoid unbalance of these induced voltages causing a voltage around the pilot loop.

Where open cambric-covered pilots are used (generally abroad), frequent transposing is necessary. A high degree of insulation is required, and for overhead pilot earth wires special types have been manufactured with both magnetic and electrostatic screening in order to reduce the induced voltage. The induced voltage depends upon the length and construction of the line and the value of earth fault current, so that it is usually expressed in V/A/km for earth fault conditions. A value of 1·9 is generally used for pilots without special screening. Even with this special screening the induced voltages may be relatively high (0·6), for example a 16 km length line with an earth-fault current of 5 kA would give an induced voltage of 5 kV. The value would be 15 kV for the unscreened type. Relay equipment connected to such pilots will thus require either isolating transformer coupling or an adequate level of insulation. In the latter case adequate labelling warning that the relay equipment and pilots are subject to high induced voltages is advisable, particularly if this equipment is mounted at the relay panel.

(c) *Telephone type pilots:* For overhead lines, the provision of special pilot cables for protection may often be uneconomic and underground telephone cables are frequently used instead. These may be privately owned by the supply company and form part of their general communication and control network or they may be rented. The type of cable is similar in both cases and presents problems in the design of suitable protection. Where pilots are rented, certain additional difficulties exist because of the limitations of voltage and current which are laid down. In addition inductive loading coils included for communication reasons create difficulties. The problems in the use of telephone pilots, particularly those of the rented type, have resulted in the design of special protective systems which are described later. More recently, the growing nonavailability of solid copper-pilot channels has created further problems which have resulted in the development of phase comparison systems of protection, using voice frequency carriers, which can be applied to a variety of telecommunication company channels. This is described in Section 10.8.9.

10.8.6   Typical systems for privately owned pilots

There are a number of forms of pilot-wire protection in use and it is not possible in this chapter to cover every type. A representative range will serve to illustrate the problems involved and their solutions. Some of these systems were designed many years ago and are still in use. In general, these are suitable for use with 7/0·67 pilots with their relatively high insulation level, whereas the more recently developed systems or modifications of systems are often suitable for use with privately owned telephone pilots.

(a) *Translay system, Type H (GEC Measurements):* This is a well known system which has been used for many years and installations will be found in many parts of the world. In its basic form it is suitable for 7/0·67 pilots, although some modified arrangements have been used with privately owned telephone pilots. It is a balanced voltage system with the addition of a directional feature which enhances the discriminating factor. The other feature which is worth noting is that at higher currents, the nonlinearity of the iron circuits causes it to become a phase-comparison system.

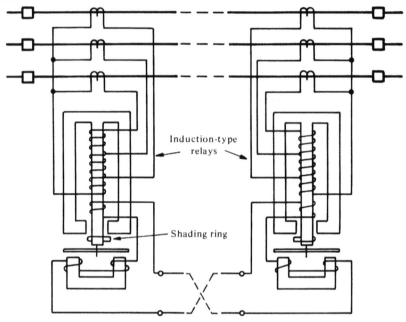

Fig. 10.8.6A   *Translay system, type H*
                   (GEC Measurements Ltd)

The general arrangement is as shown in Fig. 10.8.6A, the terminal equipment consisting of a special wattmetrical directional relay. Summation of secondary line currents is effected in the top winding of the upper magnet. As this winding is coupled to an output winding which in turn is effectively across the pilots, the upper winding can be considered as acting as a summation current transformer. This upper magnet has an air gap, and it can be considered therefore that the output voltage has the characteristics of a voltage source; the upper magnet therefore converts from a current source to a voltage source. In addition, the flux in the upper magnet will be related to the voltage on the output winding and so to the pilot voltage. This flux, which is one of the applied quantities on the relay, is therefore related to the pilot voltage. The lower magnet is energised by a winding which is connected in series with the pilots, and since the flux in this coil is the second quantity applied to the relay, we have a relay which is effectively responsive to the pilot voltage and the pilot current.

This type of relay develops a torque which is proportional to the product of the two fluxes and the sine of the angle between them. It therefore develops a maximum torque when the fluxes are at 90° and zero torque when they are in phase. In this particular application, the fluxes would be in phase if the pilot impedance were purely capacitive, and the relay would be insensitive to pilot capacitance current flowing under external faults. The pilot impedance will be generally resistive on internal faults so that a positive tripping torque would be developed provided both fluxes existed, that is fault current existed at the relay terminal concerned.

The action of this relay may be considered in terms of the complex impedance diagram developed for various fault conditions in Section 10.8.4(b). In plotting the tripping zone of a directional relay, such as the Translay relay, on this diagram (Fig. 10.8.6B) it will be seen that it generally includes those impedance vectors occurring under internal faults but would exclude that impedance occurring on external faults.

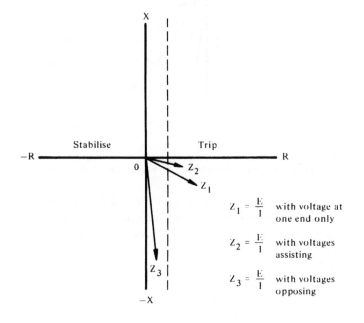

Fig. 10.8.6B *Simplified directional impedance characteristic of Translay H system*

For a practical scheme, there would be conditions where the errors of current transformers could produce unbalance in the pilots and consequently tripping tendencies on the relays. It should be remembered that the existence of one flux alone does not theoretically produce any torque on the relay. To give the relay some of the qualities of a biased differential system, a shading ring is included on the upper magnet which enables the relay to develop a restraining torque due to the presence of the upper flux alone. This shading ring has an effect on the complex impedance representation of the relay which is to distort the basic linear polar characteristic of the directional relay, as shown in Fig. 10.8.6C. This enhances the ability of the relay to distinguish between the impedance factors for internal and external faults as well as providing this bias feature. Another practical modification

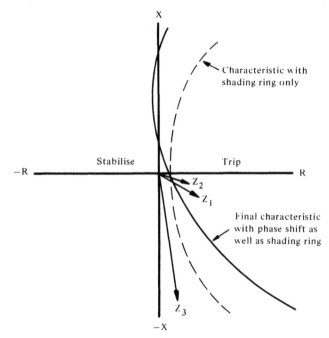

Fig. 10.8.6C   *Curvature of relay characteristic produced by shading ring in Translay H protection*

to the relay which is used to further improve the stability on the external fault impedance factor is to shift the angle of maximum torque slightly as shown, thus increasing the separation between Z_3 and the characteristic curve and the tripping zone of the relay.

The performance of this system is generally as follows

Fault settings

Earth fault	22–40%
Phase fault	45–90%
Three-phase fault	52%

Speed of operation is about 6–7 cycles at five times fault setting. Being an induction pattern relay, the times vary somewhat inversely with the fault current.

It is generally suitable for 7/0·67 pilots with 2 kV insulation level and can cope with pilot inter-core capacitance of up to about 3 μF, corresponding to a pilot loop resistance of about 300 Ω. The scheme is also suitable for 3/0·67 pilots provided the pilot loop resistance does not exceed 1000 Ω. Because of the product nature of the relay, it is a single-end tripping system, that is, only the end carrying fault current is tripped on an internal fault. In cases where it is necessary to trip both ends, d.c. intertripping relays on the same pilots may be used.

(*b*)   *Solkor A system (Reyrolle):* This is another system which has been successfully applied for a number of years. It is essentially a voltage-balance system

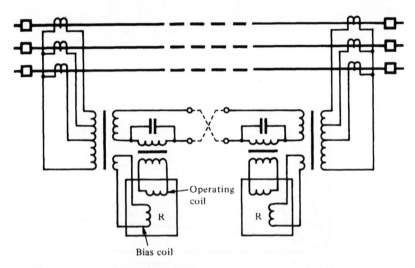

Fig. 10.8.6D   *Solkor A system (Reyrolle)*

intended for use with 7/0·67 pilots. The general arrangement is shown in Fig. 10.8.6D, from which it can be seen that the line current transformers are summed in a solid core summation transformer, the output of which energises the pilots in a voltage-balance sense.

The design of this summation transformer is such as to deliberately introduce saturation of the core at about 1·5 times the c.t. secondary rating for an earth fault energising the whole summation transformer primary winding. This use of a saturated characteristic requires accurate standardisation of the summation transformers during manufacture. The relays are of the rotary moving iron type, a.c. operated and of high sensitivity, for example 22 mVA. They are of the biased type, the bias winding being in effect energised by the pilot voltage through a winding on the summation transformer. The operating winding is fed from a relay transformer, in series with the pilots, which is shunted by a capacitor. At fault currents below saturation, the summation transformer output is sinusoidal and the system is purely differential, taking account of both the magnitude and the phase angle of the primary currents at the two ends of the feeder. At high currents when saturation takes place, the output voltage waveform is distorted and peaky as previously described, and comparison is mainly on the basis of phase angle. Because of the peaky output voltage the effect of pilot capacitance current on the relay would be exaggerated, and the shunt capacitance is used to tune the operating circuit to the fundamental frequency component. This component, of course, reaches a limiting value and does not increase, and is related to the phase angle of the primary current in a similar manner at both ends on an external fault. With such an arrangement at the high current levels we have a stabilising angle of the order of ±60°. This arrangement is very effective in that it enables fairly high stability limits to be achieved with relatively low performance current transformers, provided their characteristics

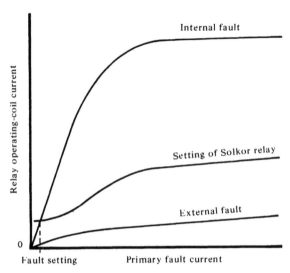

Fig. 10.8.6E   *Discrimination characteristics of Solkor A protection*

and loadings are within acceptable limits.

The limitation of fundamental pilot voltage also limits the maximum value of equivalent pilot capacitance current flowing in the relay under external faults so that, with the addition of a moderate degree of bias, maloperation can be prevented. The typical characteristics, which are shown in Fig. 10.8.6E, indicate the order of discriminating factors obtainable. The discriminating factors in practice are enhanced somewhat from the theoretical values because the tuned circuit is resonant at the higher internal fault levels and gives an additional output of current to operate the relay under internal faults. The general performance is as follows

Fault settings
Earth fault settings	40–60% (normal)
or	10–20%
Phase-phase faults	120–240%
Three-phase fault	140%

Operating time at five times fault setting is about 6 cycles. The system is suitable for 7/0·67 pilot loops up to 400 Ω. A modification to this scheme has been brought out recently to overcome some of the problems of mechanical instability experienced in some installations with the early design of rotary sensitive relay. In this arrangement a transductor relay is used in place of the rotary sensitive relay, as shown in Fig. 10.8.6F. Both the electrical and mechanical stability are improved but the remainder of the performance is about the same.

(c)   *Solkor R system (Reyrolle Protection):* This system was adapted from a more elaborate system developed for use with rented pilots. By taking advantage of the easier requirements of privately owned telephone pilots, a much simpler form of

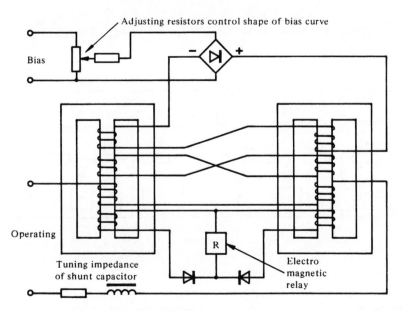

Fig. 10.8.6F *Transductor relay for Solkor A protection*

protection was obtained. However, the basic principles are the same and these will be explained here rather than under the rented-pilot type system. The method of operation may be explained in two different ways. First, the system can be considered as a current-balance type with the relays shunt-connected at the pilot terminals. The problem of imparting differential characteristics and obtaining correct relaying connections for such an arrangement has been overcome by introducing half-wave rectifiers into the pilot loop and the relay connection, as shown in Fig. 10.8.6G.

The pilot loop contains a half-wave rectifier at each end shunted by a resistance R_a and so arranged that conduction is in opposite directions at each end. The relays are connected across the pilot terminals in series with the half-wave rectifiers as shown. Taking the external fault condition and considering successive half cycles we can see that if $R_a = R_p$, then the equipotential point on the pilots will alternate

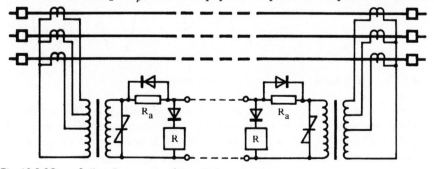

Fig. 10.8.6G *Solkor R protection (Reyrolle)*

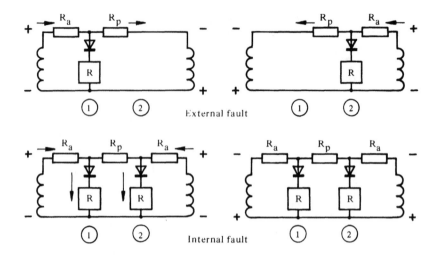

External fault

Internal fault

Voltages at relaying points 1 and 2

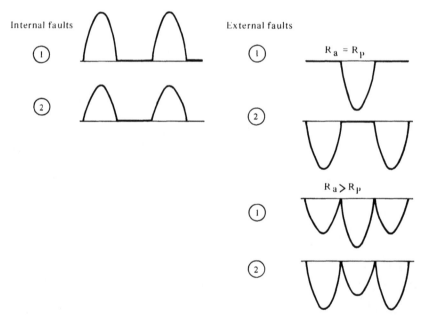

Internal faults

External faults

$R_a = R_p$

$R_a > R_p$

Fig. 10.8.6H   *Principle of operation of Solkor R protection*

between ends, one relay being at the midpoint on each half cycle. The rectifier in series with the relay circuit is so arranged that it prevents current flow under the polarity of voltage that the relay experiences when it is not at the midpoint. In this way each relay effectively experiences a midpoint connection.

An alternative description of this action is that the rectifiers act as switches and permit the use of a single pilot loop for a two-way signalling action using a system of time-sharing switched at the power frequency.

In a practical scheme, the resistor R_a is made greater than R_p and so on external faults causes reverse voltage on the relay rectifier to occur on both half-cycles, as shown in Fig. 10.8.6H. This effectively displaces the midpoint connection further towards the relaying end as shown. This particular feature is equivalent to applying a bias signal to the relay, and effectively stabilises the relay on through-fault conditions even with considerable transient errors in the current transformers. In most practical cases R_a can be made a pure resistance, and it is not necessary to mimic the pilot capacitance. The pilot voltage is unidirectional on external faults, although the current from the summation transformers is alternating. This feature, together with offsetting of the midpoint, has the effect of reducing the influence of pilot capacitance on through-fault stability during external faults.

Under internal-fault conditions, the relay is fed by half-wave pulses of current, although for a single-end fault the distribution between ends is such that only the relay at the feeding end can be considered operative. The system is therefore of the single-end tripping type. In order to limit the peak voltage, Metrosil devices are included across the output of the summation transformers. These limit the pilot voltage to about 350 V and give the system a phase-comparison action. With these voltage levels and a maximum loop impedance of 1000 Ω, it is possible to use a simple attracted-armature relay with adequate contact performance for direct tripping of circuit breakers. The summation transformers are insulated to a level of 5 kV to cater for applications where there might be longitudinal induction.

Fault settings

Earth fault	R = 25% Y = 32% B = 42%
Phase-phase faults	R-Y = 125% Y-B = 125%
	B-R = 62%
Three-phase fault	72%

Operating time is about 3 cycles at three times setting. The system is suitable for use with 7/0·67 pilots up to 30 km, 5·68 kg/km telephone pilots up to 15 km and 11·36 kg/km telephone pilots up to 30 km.

(*d*) *Type DSB7 biased differential protection (GEC Measurements):* This system is based on the voltage balance principle and, as it is primarily intended for application to teed feeders, the derived quantities at each relaying point are not limited, but vary directly with the associated fault current. Fig. 10.8.6I shows a simplified arrangement, two pilot loops being formed by a 4-core pilot, one loop summating vectorially the derived voltages of the relaying points for the operating quantity, and the other loop summating arithmetically these voltages to form the biasing quantity. The derived voltages are obtained from tapped air-gapped (quadrature) summation transformers at each relaying point.

The principle of operation is that under normal load or through fault conditions the summated bias quantity greatly exceeds the summated operating quantity,

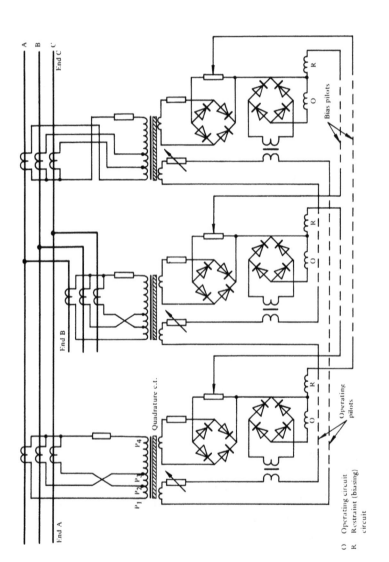

Fig. 10.8.6I   *Type DSB7 protection applied to a three-ended feed (GEC Measurements Ltd)*

which is theoretically zero. Under internal fault conditions, the operating quantity at each end will exceed the biasing quantity with an adequate margin. The operating and bias quantities are compared at each end by means of two windings on a permanent magnet moving coil relay.

The particular requirements for a system of this type applied to teed feeders are:

(i) The derived quantities should be linearly related to the primary current, and, on through faults, balance will be required for differing values of fault current at the various ends. In this scheme the quadrature summation transformers are linear up to thirty times rated current.

(ii) The biasing level should be adequate to overcome the real errors in vectorial summation on through faults. With a linear system, this imbalance, for example, due to pilot capitance, increases with fault current.

(iii) Perfect balance between current transformers operating at differing values of primary current is not achievable, particularly under transient conditions. The biasing level needs to be capable of stabilising against operation under such conditions of through fault. This should also take account of differing designs of current transformers at each end.

Additional stabilising features are included in the complete scheme. Two second-harmonic tuned circuits are included in the operating circuits at each end, one to filter the harmonic from the operating coil and the other to provide an additional bias related to this harmonic. This feature enhances stability on magnetising inrush currents and also on transient distortion of current transformers.

A further feature is the inclusion of a low-impedance stabilising resistor in the star connection of the current transformer secondaries.

The following figures give typical performance data:

Fault settings	Summation transformer tappings give nominal settings for earth fault between 20-33%, for phase faults between 50-100% and three-phase settings of 58%.
Operating times	from 60-80 ms at ten times the fault setting.
Stability	This is nominally thirty times rated current but achieved values depend on pilot length and characteristics. This may reduce to about 10-15 times for pilots of 4 mF capacity and 1200 Ω loop resistance. Note, this is acceptable in most cases as the fault current reduces accordingly with increasing feeder length.

(*e*)   *Translay S (GEC Measurements):* This protection is based on the circulating current principle and uses static phase comparators as the measuring elements. The basic circuit arrangement in Fig. 10.8.6J shows a summation current transformer T1 at each line end the neutral section of which is tapped to provide alternative sensitivities for earth faults.

The secondary winding supplies current to the relay and the pilot circuit in parallel with a nonlinear resistor, RVD, which is nonconducting at load current

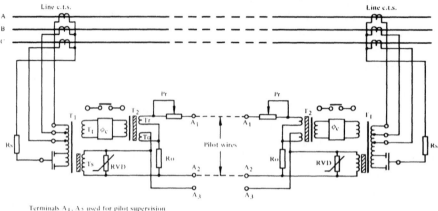

Terminals A₁, A₂ used for pilot supervision
Terminals A₁, A₃ used for unstabilising and intertripping

Fig. 10.8.6J   Translay S protection (GEC Measurements Ltd.)

levels and which, under heavy fault conditions, conducts an increasing current thereby limiting the maximum secondary voltage. At normal current levels the secondary current flows through the operate winding T_o on transformer T2 and then divides into two separate paths, one through resistor R_o and the other through the restraint winding T_r of T2, the pilot circuit and resistor R_o of the remote relay.

The currents in windings T_o and T_r are summated and fed to the phase comparator to be compared with the voltage across winding T_t of transformer T1. The voltage across T_t is in phase with that across the secondary winding T_s which in turn is substantially the voltage across R_o.

Taking into account the relative values of winding ratios and circuit resistance values, inputs to the phase comparator are:

$$(I_A + 2I_B) \text{ and } (2I_A + I_B)$$

where I_A and I_B are the currents fed into each line end and for through faults $I_A = -I_B$. By putting each input in turn to zero we find the values of I_B in terms of I_A for which the systems is stable. Thus,

$$I_A + 2I_B = 0 \text{ and } I_B = -I_A/2$$

$$2I_A + I_B = 0 \text{ and } I_B = -2I_A$$

so that stability is obtained for values of I_B outside the above limits because the inputs to the phase comparator will be of opposite sign.

The phase comparator has angular limits of $\pm 90°$ giving a circular bias character-

Fig. 10.8.6K *Translay S protection (GEC Measurements Ltd.)*

istic in the complex current plane and in fact the comparator is equivalent to the traditional low impedance biased relays which are often identified with differential protection. Such relays have an amplitude criterion for operation such that

$$|I_A + I_B| \geqslant K |I_A - I_B|$$

and the equivalent phase comparator inputs are

$$I_A(1-K) + I_B(1+K) \text{ and } I_A (1+K) + I_B(1-K)$$

By putting $K = {}^1/_3$, the phase comparator inputs become

$$\tfrac{2}{3} (I_A + 2I_B) \text{ and } \tfrac{2}{3} (2I_A + I_B)$$

i.e. of the form used in Translay S.

The input circuits of the phase comparator are tuned to the power frequency so that the threshold of operation increases with frequency. This desensitises the relay to the transient high-frequency charging current that flows into the line when it is energised and also reduces any distortion caused by saturation of the current transformers.

In order to maintain the bias characteristic at the designed value it is necessary to pad the pilot loop resistance to 1000 Ω and a padding resistor P_r is provided in the relay for this purpose. However, when pilot isolation transformers are used the range of primary taps enables pilots of loop resistance up to 2500 Ω to be matched to the relay. The equipment is illustrated in Fig. 10.8.6K.

The following gives typical performance data:

Fault settings	Summation transformer tappings give nominal settings for earth faults between 12-33%, for phase faults between 8-44% and three-phase setting of 51%. These settings may be further varied between limits of 0·5 - 2·0 by a setting multiplier.
Operating time	This is adjustable by a special factor K_t so that the operating time can be increased with a corresponding decrease in knee-point voltage requirements for the c.t.s.

K_t 40 20 14 6
Average time at 5 x
Setting current (ms) 30 50 65 90

10.8.7   Use of rented pilots

The significant application of rented pilots to feeder protection in the UK dates from the post-war expansion of the transmission networks at 132 kV. At this period the main protection for such networks had become phase-comparison carrier-current protection which was both complex and expensive. Many cases existed where the length of the overhead line was 50 km or less and in these cases it was not felt that carrier protection was justified. Manufacturers of protective gear were asked to investigate the possibility of extending pilot-wire protection to the use of rented pilots. Most of the problems which have already been discussed are emphasised in the case of rented pilots but, in addition, particular problems are introduced by the conditions of renting from the telephone authority.

At this point it should be noted that it is feasible to apply differential principles to extreme pilot conditions provided the protection is accurately matched to (and compensated for) the pilot characteristics. This, however, was not the preferred solution at the time as the need was for a standard form of protection which required the minimum of adjustment and had the minimum of dependence on the pilot characteristics. Provided the many technical problems can be overcome, the application of such systems is economically attractive compared with the alternative forms of protection using distance relays or carrier-current equipment.

Before considering the particular systems resulting from this development it is useful to review the special problems associated with the use of telephone pilots as they are common to all designs.

(*a*) *Discriminating factor:* With the lengths of pilots involved, the characteristics of high loop resistance and relatively low shunt capacitive impedances cause a problem in respect of basic discriminating factor. This has already been referred to but in this type of application it is particularly shown up by the fact that a 30 km, 5·68 kg/km cable gives a discriminating factor of two (without special compensation) this being reduced to 1·3 if the length is increased to 50 km. The acceptance of the basic discriminating factor, which is common with 7/0·67 pilots, is no longer possible with the lengths of rented pilots required, and additional design features must be included to enhance this factor. There are a number of different ways of doing this all of which involve some complication, limitation or loss of flexibility in the overall equipment. Some of the methods which may be used are as follows.

(i) *Tuning of pilot capacitance:* With a voltage-balance arrangement, the effect of pilot capacitance current on external faults may be reduced by connecting a chosen value of inductance across the terminals of each end of the pilots, as shown in Fig. 10.8.7A. This inductance would normally be tuned to half the pilot capacitance, the arrangement being symmetrical. The discriminating factor can be improved up to about 6 by this technique, the distributed nature of the pilot resistance and capacitance limiting the sharpness of tuning which may be achieved.

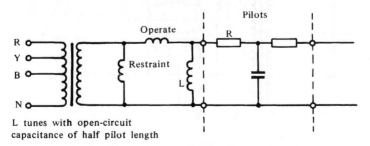

L tunes with open-circuit
capacitance of half pilot length

Fig. 10.8.7A *Tuning of pilot circuit capacitance*

In any case, it is necessary to limit this so that the effect of system frequency variations should not be pronounced. This improvement of discriminating factor is further limited because in most systems the use of limiting produces nonsinusoidal waveforms and such compensation can only be effective at the fundamental frequency. This method has been successfully used but it involves a compensation which must be suited to the particular type of pilot and route length. If the pilot is changed then the compensation must be altered accordingly.

(ii) *Use of complex-impedance relay characteristics:* As previously described, improvements in discriminating factor can be achieved if account is taken of the phase angle of the effective pilot impedance as well as its magnitude. This technique has already been used in the Translay H system with respect to 7/0·67 pilots, and this system was developed in a modified form to suit telephone pilots up to pilot loops of 1000 Ω. With the considerable developments in the distance relay field and sensitive relays with a variety of complex-impedance characteristics, it is possible to adapt such relays to pilot-wire protection.

(iii)   *Compensation for pilot capacitance current:* More exact ways of compensating for pilot capacitance current are possible compared with the simple tuning technique referred to above. The technique is generally to create a replica impedance corresponding to the pilot under external fault conditions. Such impedances are energised by the pilot voltage or some voltage corresponding to this, and the current through them is used to counteract the pilot capacitance current or create more exact balance points. This method does not involve tuning but again is dependent on the type and length of pilots. One possible arrangement is shown in Fig. 10.8.7B.

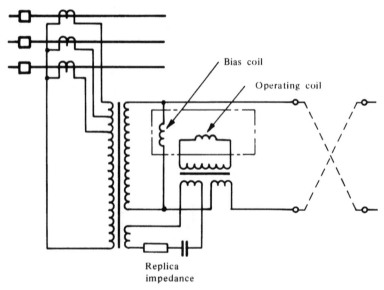

Fig. 10.8.7B   *Pilot compensation by replica impedance*

(iv)   *Use of current balance:* As previously noted, the basic discriminating factor of a current-balance arrangement can be fairly high if the problem of correct relaying connection is solved. There are methods of overcoming this problem without resorting to a three-core pilot. One method which is shown in Fig. 10.8.7C uses a replica impedance at each end which is arranged to form one arm of a bridge, the other arm being one half of the pilot loop. Under external fault conditions, the pilots may be considered as short-circuited at the midpoint and each end finishes as a balanced bridge, the relay being theoretically unenergised. A bias signal may be obtained from the circulating current to ensure stability under practical conditions of external faults. Under internal faults, each relay is energised equally, even when

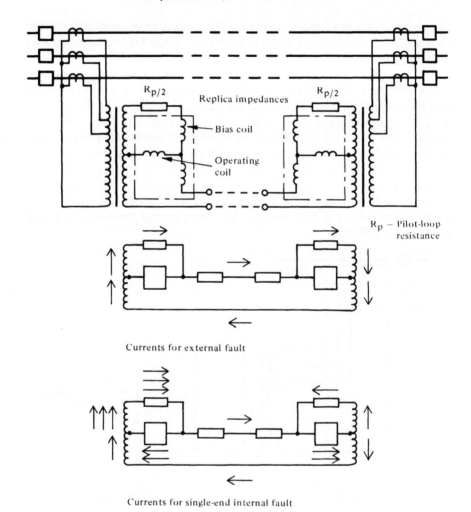

$R_{p/2}$ Replica impedances $R_{p/2}$

— Bias coil

— Operating coil

R_p — Pilot-loop resistance

Currents for external fault

Currents for single-end internal fault

Fig. 10.8.7C *Pilot compensation in current balance system by replica impedance. Replica impedance set to impedance of short-circuited half pilot section*

the fault is fed from one end only. This system requires the adjustment of the replica impedances to suit the type and length of pilot.

(*b*)   *Regulations affecting the use of telephone type pilots:*

(i)   *Pilot voltage and current:* The conditions governing the use of telephone type pilots in power systems are particularly stringent, and when the use of such pilots was considered for protection these conditions had to be interpreted and eased by negotiation with the telephone authority. The conditions generally related to safety aspects and external interference rather than the electrical capabilities of the pilot. At present the regulations require that the peak voltage developed on the pilots under maximum fault-current conditions shall not exceed 130 V and that the pilot current shall not exceed 60 mA r.m.s. These values are in excess of the normal values applied to Post Office circuits and were only permitted because of the short duration of fault conditions.

(ii)   *Insulation and protective gaps:* The present regulations require all equipment connected to the pilots to be suitably insulated from power-system equipment and circuits, for example current transformer, d.c. and a.c. auxiliary circuits. The level of this insulation is 15 kV r.m.s. and screens (Fig. 10.8.7D) must be provided. This is a high level of insulation when considered in relation to the design of relays and auxiliary current transformers, and has greatly contributed to the size and complexity of the overall equipment. Protective gaps on the protective gear equipment are also required and these are usually of the glow-discharge type with a setting of about 750 V. The introduction of these gaps does not affect the protection provided they do not fire under normal conditions (including system fault conditions).

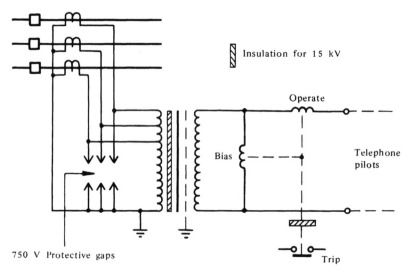

Fig. 10.8.7D   *Use of insulation and protective gaps on system using telephone pilots*

(c)   *Relays power and limiting:* The values of permitted maximum voltage and current on the pilots, taken in conjunction with high loop impedances of 2000 Ω or more, represent a serious limitation to the power which it is possible to transmit over the pilots. Taking the maximum peak voltage of 130 V and sinusoidal waveform, the maximum power possible in the relays for a voltage-balance system would be about 4 W (in the local relay) for fault current fed at one end only. This would be for optimum impedance termination which in practice may not be possible. The actual power in the relay would be nearer 0·5 W and this would be at the maximum level of fault current. With a linear system where the voltages are proportional to the fault current, this would give a power of the order of 0·2 mW at the relay setting, assuming an earth-fault setting of 50% and a maximum earth-

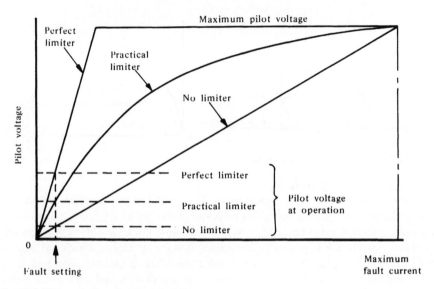

Fig. 10.8.7E *Action of voltage limiting*

fault current of 2500 %. This is a very sensitive relay, bearing in mind the contact requirements and the insulation levels required. The general action of limiting is shown in Fig. 10.8.7E, the most convenient form at present being the non-linear resistor (Metrosil) which has a voltage/current characteristic defined by $V = CI\beta$ where β is an index of about 0·2-0·3 and C is a constant. The limiting action of such a device may be improved by the inclusion of a hybrid transformer and a resistor, as shown in Fig. 10.8.7F, or by a multistage arrangement as in Fig. 10.8.7G.

The introduction of such limiting defines that the protective systems are basically phase-comparison over the pilots. Zener diodes are also used for voltage

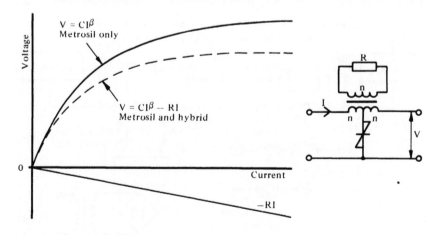

Fig. 10.8.7F *Improvement to limiting using hybrid transformer*

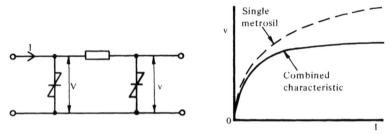

Fig. 10.8.7G   *Two-stage voltage limiter*

limiting in modern equipments.

(*d*)   *Requirements for pilot supervision:* The pilot link is essential for correct discrimination. In systems using buried pilots of the 7/0·67 type, the reliability of the pilot system is rated fairly high. When telecommunication company pilots are used for protection it is generally considered that the reliability is reduced, due to the possibility of unauthorised interference, since these circuits form part of a communication system and will go through junction boxes and exchanges. Even though special precautions are taken, pilots may be short-circuited, open-circuited or reversed, and it is becoming standard practice to provide supervision equipment so that these conditions may be detected and an alarm given. In this respect, careful consideration should be given to the three-phase fault setting under the faulted pilot conditions mentioned. It is not possible to prevent maloperation of the protection under through-fault conditions with such pilot voltages, but it is possible to keep the corresponding three-phase settings above maximum load rating. Most supervision schemes are basically the same and involve the circulation of a d.c. current of about 5 mA round the pilot loop as shown in Fig. 10.8.7H.

A polarised relay at the receiving end of the pilots is operated by the supervision current. Any of the faulty pilot conditions of short-circuit, open circuit or

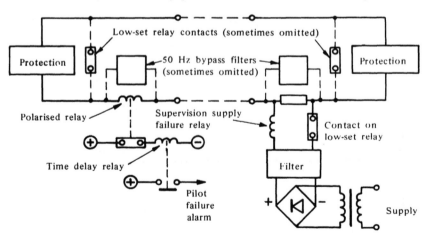

Fig. 10.8.7H   *Typical arrangement for pilot supervision*

reversal will cause the supervision relay to reset and an alarm to be given. A monitor relay at the sending end is arranged to detect the failure of supervision supply and give an alarm.

The supervision supply is disconnected during fault conditions so that the supervision current does not affect the discriminating action of the protection. All alarms are provided with time delays of three seconds or more so that alarms are not given due to disconnection under primary fault conditions. The supervision supply is usually provided from an auxiliary a.c. supply, through rectifiers and filters. The Telephone authority regulations require a high degree of smoothing in order to keep a low level of interference with adjacent communication circuits.

(*e*)   *Requirements for starting relays:* The philosophy of starting relays has been explained previously and it is only necessary here to state that they are prime requirements because of the supervision aspect. The provision of starting relays which have close settings, particularly for the three-phase condition, is very difficult and this is particularly emphasised in the case of two-stage starting. Under all conditions the starting relay settings must exceed those of the discriminating link, so that there is no condition of starting relays being operated without a corresponding adequacy of sensitivity on the discriminating link. The starting equipment may consist of separate relays energised by the secondary line and residual currents or, alternatively, a separate summation network without output relays. In both cases the relays must have a high reset ratio, a high operating speed and equality or near equality of phase and three-phase fault settings. The earth-fault sensitivity is largely independent of the phase-fault sensitivity and does not constitute a problem.

(*f*)   *Pilot circuit characteristics:* The pilot link in the case of telecommunication company pilots is liable to rerouting when the necessity arises. In order to facilitate design and application, it is normal to use a standard pilot-loop resistance. Where the actual pilot resistance is below this value the loop is padded up by the addition of resistances at each end so that the effective loop is always the same. The pilot capacitance will vary with the route, and where pilot tuning is used, this must be adjusted as well as the pilot padding resistors.

The upper value of 3500 Ω based on a conductor of, say, 5·68 kg/km makes discrimination difficult even with the additional compensations which are used. In specifying the performance of protection in terms of pilots it is common to express the upper value of pilot-loop impedance and total pilot capacitance. This value of capacitance will generally be lower than the equivalent length of pilot defined by the maximum loop resistance. The 3500 Ω figure is based on the inclusion of lengths of 2·5 kg/km aerial conductor at the terminations, for example, between the substations and the nearest exchange and, of course, the aerial conductors have negligible capacitance compared with the cables.

10.8.8   Typical systems for use with rented pilots

Two types of protection are described, these represent widely differing solutions to

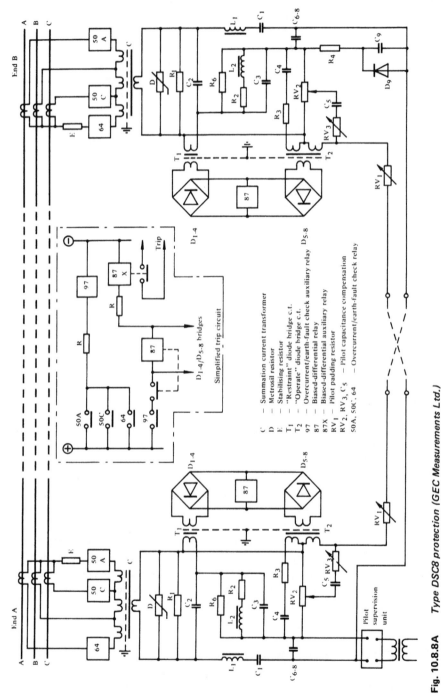

Fig. 10.8.8A   *Type DSC8 protection (GEC Measurements Ltd.)*

the problems involved and therefore make an interesting comparison.

(a)   *Type DSC8 (GEC Measurements):* This is a voltage balance system in which the relays respond to the apparent impedance of the pilot circuit. The general arrangement is as shown in Fig. 10.8.8A. Operation of the protection may be considered in two parts, the 'restraint' circuit and the 'operate' circuit.

The 'restraint' circuit comprises a shunt connection of R6, C_{6-8} and (a) $R_2 L_2$ at frequencies below fundamental and (b) C_3 at frequencies above fundamental. The current derived from the voltage developed across R_1 is transformed by auxiliary transformer T1 and rectified by the diode bridge (D1-4) to produce the restraint component of the d.c. current fed to the differential relay (87). Circuit $L_1 C_1$ is tuned to fundamental frequency to reduce the effect of harmonics on the restraint, operate and pilot circuits. Capacitor C_2 provides additional restraint during transient conditions occurring at the inception of through faults, or their clearance, when pilot capacity discharge current may flow.

The 'operate' circuit comprises essentially, two circuits in opposition: the half-winding of T2 in series with the pilots and the pilot loading resistor R_{v1} and the other half winding of T_2 in series with $R_3 C_4$ (representing a lumped, constant equivalent of half the maximum pilot shunt capacitance). The pilot shunt circuit $C_5 R_{v3}$ allows compensation for varying values of pilot capacitance and R_{v2} allows compensation for the resistance of varying lengths of pilot and the combinations serves to improve the discriminating factor. The pilot padding resistor R_{v1} is adjusted to half the difference between the minimum permissible pilot loop resistance of 3000 Ω and the actual pilot loop resistance in use. The output current from T_2 drives the 'operate' diode bridge (D5-8) and is approximately zero when the pilot compensation has been correctly set.

The d.c. outputs of the two diode bridges summate algebraically and the resultant current is measured on the moving coil unit (87).

Under through-fault conditions, ends A and B present voltages of opposite sign to their pilot ends, but since the pilots are crossconnected, only a small value current is caused to flow in the apparently high-impedance pilot loop. The much larger value of current circulated through T_1, therefore, restrains the differential relay.

Under internal-fault conditions, the primary system current at one end is reversed in direction and the pilot terminal voltages are then such as to drive current around the apparently low impedance pilot loop. The d.c. current in T_2 is then greater than that in the diode bridge of T_1 and the difference current circulates through the differential relay to cause operation.

Typical protection performance characteristics are shown in Fig. 10.8.8B.

The pilot supervision unit injects a small d.c. current into the pilot loop. At the end remote from the supervision unit, the circuit comprising R_4 and D_9 provides a low-resistance path to the supervising d.c. current, capacitors C6-8 shunting the a.c. in the pilot circuit. The supervision unit provides alarms for 'pilot circuit faulty' and 'supervision supply fail' conditions.

The tapped summation transformer (C) provides the 15 kV insulation required to comply with Telephone authority requirements and in conjunction with resistor

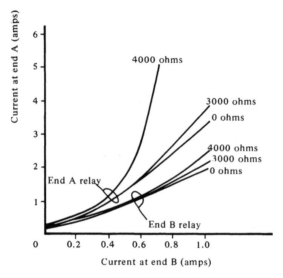

Pilots: 5.68 kg loaded
End A current in phase with end B

(a) Bias characteristic with
varying pilot resistance

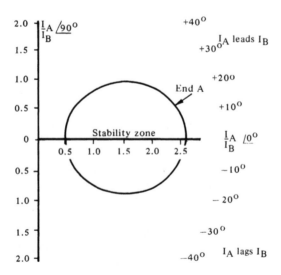

Pilots: 2000 ohms 5.68 kg/km loaded

(b) Polar characteristics

Fig. 10.8.8B   *Characteristics of type DSC8 protection*

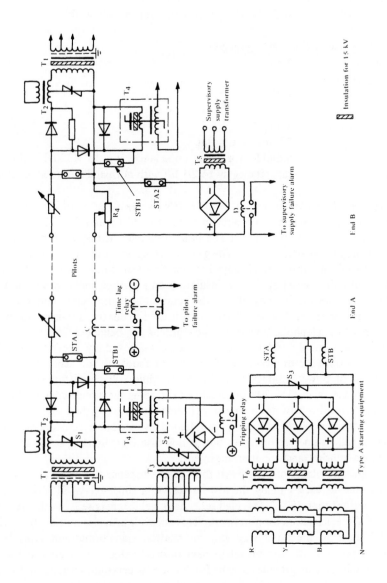

Fig. 10.8.8C *High-resistance pilot wire protection (Reyrolle)*

R_1 develops voltages for the restraint, operate and pilot circuits. The voltage applied to the pilots is limited to 130 V (peak) by the shunt metrosil (D). To ensure that tripping cannot occur because of defective pilots, operation of the differential relay auxiliary tripping relay (87X) is controlled by overcurrent and earth-fault check relays (50A, 50C and 64). The performance is generally as follows

Fault Settings (including starting relays)

Earth faults	40%
Phase-to-phase and three-phase faults	130%
Operating time at 5 x fault setting	2-3 cycles
Stability	30 times c.t. rating
Pilots	Suitable for loop impedances up to 4000Ω with total capacitance up to 2·1 μF with normal distributed loading.

(*b*)   *Reyrolle high-resistance pilot wire (HRPW) scheme.* Some of the advantages of a current-balance system have already been referred to, and this principle forms the basis of a rented pilot scheme. The basic principles of operation have already been explained in Section 10.8.6C. The general arrangement of the scheme is shown in Fig. 10.8.8C. A conventional tapped summation transformer is used, this being provided with the necessary 15 kV insulation and earthed metal screen between windings. As previously described, the current in the relay connection is unidirectional on internal faults. This current is used to control a transductor which is in series with a telephone-type relay, the combination being fed from an auxiliary constant-voltage supply obtained through a network fed by the c.t. secondary currents. This transductor gives a switching gain of about 20, and helps to overcome the problem of power limitation of the pilots while still permitting the use of a relatively robust relay (15—20 mW) capable of tripping the circuit breaker directly. The transductor also provides a convenient means of isolating the relay circuits and contacts from the pilots. Limiting is obtained by nonlinear resistors across the summation transformer secondaries, but in this case their limiting action is supplemented by a hybrid transformer arrangement.

There are two stages of starting in this scheme because the pilots are normally short-circuited at each end by the low-set relays. A single-relay (STA) is used for low-set and another (STB) for high-set, these being energised by the rectified output of a special summation network. This network provides outputs for various types of faults in the same way as a summation transformer but gives better equality of settings. The low-set relays have normally closed contacts which short-circuit the pilots. This provides a path for the d.c. supervision current, and so prevents this current from flowing in the windings of the summation transformers. This feature also has the effect of making the single-end fault settings independent of the through-load. The low-set relays have contacts which normally short-circuit the control windings of the transductors, and thus ensure that the discriminating link is completed by the operation of the low-set relays on external faults before permitting the comparison to be effective. Supervision equipment of the type previously described is included.

Typical performance figures are as follows:

Fault settings
Earth fault   50-60%
Phase fault   190%
Three-phase fault   160%
Operating time   2-3 cycles at five times setting
Stability   At least 25 times c.t. rating
Pilots   Suitable for loop resistance up to 3500 Ω and pilot capaci-
                   tance up to $2\mu F$ with standard loadings

10.8.9   V.F. phase-comparison protection (Reyrolle Protection)

This system uses phase-comparison principles and is a development of the power-line-carrier phase-comparison protection described later in Section 10.10 adapted to operate over rented Post office voice frequency channels. In designing a phase comparison system to operate over this type of communication link a number of restrictions not normally encountered with power-line-carrier phase-comparison protection need to be overcome. In the case of rented telephone channels the continuous power transmission must not exceed −13 dBm; for short durations the power may be boosted to 0 dBm. Neither the length nor the composition of the channel can be guaranteed to be constant during the period of service and may be changed by automatic route switching without prior knowledge of the user. Attenuation in the link may vary by ±3dB with time and may also vary by −1 to +4 dB over the usable frequency range. Isolation transformers in the link may also add losses of between 0·5 and 1·5 dB per transformer and the group delay over the usable frequency range may be in the order of 1 ms. As well as accommodating the characteristics of the channel the equipment must cater for the presence of random noise induced into the channel before, or at the time of, power system faults.

The main items of the v.f. phase comparison scheme are shown in Fig. 10.8.9A. At each end of the protected line a terminal equipment is situated comprising a power frequency section which interfaces with the power system and a v.f. section which is connected to the channel, e.g. a four wire v.f. communication circuit.

The basic principle of operation of the protection is the same as for power-line-carrier (p.l.c.) phase-comparison protection. In the v.f. equipment information is transmitted in the form of a frequency shift keyed signal. In this case the signal is transmitted continuously and the frequency of the signal is switched between two values in accordance with the modulation signal. The low-frequency signal would correspond to carrier transmission in the p.l.c. case and the high frequency would correspond to no carrier or gap. The continuously transmitted signal with no modulation is at the high frequency and is therefore directly analogous to the p.l.c. case. Thus, in the v.f. equipment the condition which determines tripping is the effective duration of the high-frequency signal at each end of the line after the received signal from the remote end and the local signal have been combined in an equivalent manner to the carrier blocks in the p.l.c. case.

In the p.l.c. case the carrier signal travels at almost the speed of light and the transmission delay incurred is normally accommodated in the stability angle setting of the equipment. In the v.f. case the transmission delay may exceed 5 ms and is not constant due to possible rerouting and cannot therefore be catered for within an acceptable stability angle setting. This problem is overcome in the v.f. phase comparison by measuring the total transmission delay around the channel loop and delaying the local signal by half of the loop delay before comparison with the remote signal which has itself been delayed by the communication link. In this way, even for variations in routing, the local and remote signals maintain their relative phase difference except for communication errors. As the communication errors are small and substantially constant they are catered for in the stability angle setting.

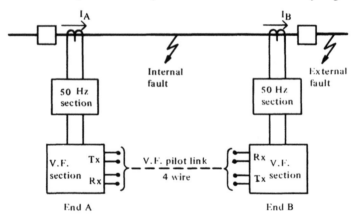

Fig. 10.8.9A   *V.F. phase comparison protection*

The principle employed in the automatic compensation for transmission delay is shown in Fig. 10.8.9B. Under healthy power system conditions a measurement pulse, consisting of a short burst of low-frequency signal, is transmitted by the master end to the slave end and is reflected by the slave. The master measures the time from transmission to reception of the pulse and stores the value. A short time after receiving the master end pulse the slave transmits a measurement pulse which is reflected by the master and the transmission time is stored by the slave. This process is repeated continuously and each end of the equipment computes an average value of the total transmission delay which is used for automatic compensation.

The main features of the v.f. section of the equipment in relation to the communication link may be summarised as follows:

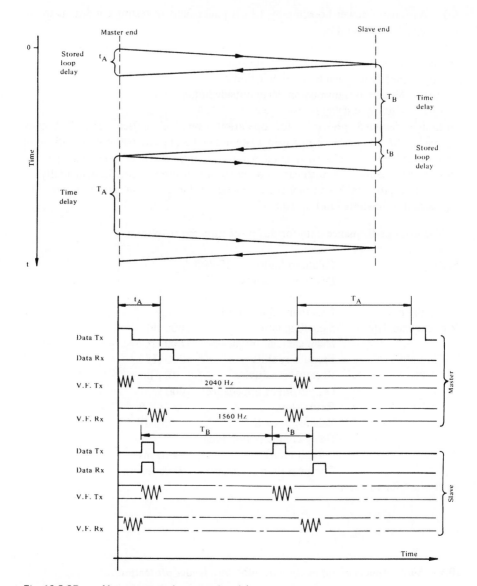

Fig. 10.8.9B *Measurement of propagation delay*

(*a*) Continuous evaluation of the transmission delay including terminal equipments.

(*b*) Automatic compensation for phase error caused by transmission delay.

(*c*) Signal quality circuits employed to detect the condition of poor signal to noise ratio and prevent the measured transmission delay from being affected by the noise.

(*d*) . Automatic power boosting by 13 dB under fault or testing conditions to improve signal to noise ratio.

(*e*)   Continuous channel monitoring, including transmitter and receiver, with alarms provided for the following conditions:

    (i)   Received signal level outside limits.

    (ii)   Measured transmission delay outside limits.

    (iii) Failure of delay measurement.

Although designed primarily for operation over Post Office voice frequency channels, the v.f. phase-comparison system can be applied without modification over other bearer circuits such as microwave or single sideband power-line carrier. In each application the automatic compensation feature is not fundamentally required but provides a comprehensive check of the communication circuit. The equipment is illustrated in Fig. 10.8.9C.

The main performance data for this protection are as follows:

Settings	Balanced faults	30-70% I_n
	Phase-phase faults	20-48% I_n
	Earth faults	36-84% I_n
Operating time	Less than 40 ms at 3 times setting	
V.F. channel data	Signalling rate	600 Bd
	Mean channel frequency	1800 Hz
	Frequency shift	±240 Hz
	Receiver sensitivity	−40 dBm
	Output/input impedance	600 Ω
	Return loss	> 14 dB
	Balance to earth	> 58 dB
	Delay compensation range	
	(half loop)	4-10 ms
	Compensation accuracy	0·05 ms

where I_n = Rated current.

10.9   Some aspects of application of pilot-wire feeder protection

10.9.1   General

In this Section some of the practical aspects which should be borne in mind when applying pilot-wire feeder protection are considered. Many factors will influence the ultimate choice of a particular scheme, such as the sensitivity requirements, stability, single-end or double-end tripping, the speed of tripping and, perhaps, the limitations which may exist in the provision of current transformers.

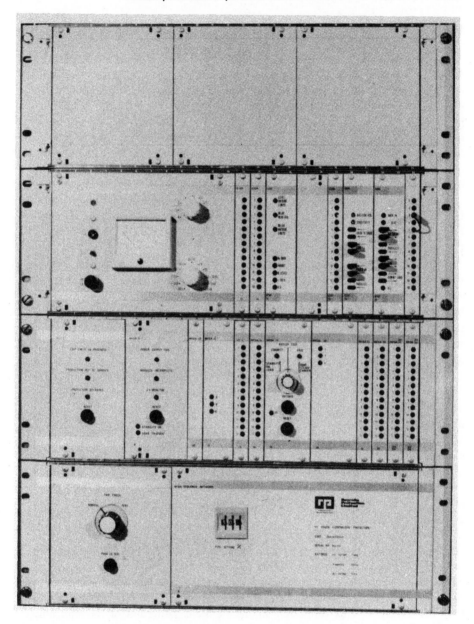

Fig. 10.8.9C *Voice frequency phase-comparison protection (Reyrolle)*

10.9.2   Current transformer requirements

In general, it is comparatively easy to design protection to operate for internal faults. It is not so easy to design for nonoperation on external faults, and pilot-wire protection is no exception in this respect. Some of the factors affecting the discriminating ability of feeder protection have already been discussed, but these have mainly been concerned with 'steady-state' conditions and have generally ignored the problems of through-fault stability which arise from the use of practical (as opposed to ideal) current transformers. The problems arising in the transient period immediately following the occurrence of a fault often play a particularly important part. It is not possible to go into the theory of transient unbalance in this Chapter, but we can say that the performance of high-speed differential protection is largely dependent upon the current transformer characteristics, the burden imposed on them, the magnitude and characteristics of the primary current and the response time of the protection.

With a two-ended network such as a feeder, a high degree of stability on through-faults may be obtained with current transformers which saturate to a considerable extent under transient conditions, provided that the current transformers are basically of the same design. It is usually necessary, however, to avoid steady-state saturation, particularly if this can occur at one end only. In assessing this steady-state saturation all factors have to be taken into account, for example the maximum through-fault current, the internal burden of the current transformers, the external lead burden, and the burden of the relay equipment. Most manufacturers provide formulas which specify these minimum current transformer requirements. These formulas are usually arrived at through extensive type-tests and will include, if the scheme is a high-speed scheme, some margin for transient conditions. In addition, it is sometimes necessary to define the maximum permissible exciting current of the current transformers in the unsaturated region in order to limit the inaccuracies at low-level currents. This is particularly so where the current transformers are of different designs. These low-level inaccuracies can represent an encroachment on steady-state stability and can also influence the effective fault settings of the protection.

Although similarity of current transformers at the two ends of the feeder is advantageous and often possible, particularly in metal-clad gear at lower voltages, it should be noted that unequal saturation under severe transient conditions (and consequently high transient unbalance) may still occur. Such a case would occur with unequal lead burdens at the two ends, resulting in different maximum prospective flux densities. This can result in the transient saturation of one end only, causing relatively large out-of-balance currents of considerable duration compared with the time necessary to produce relay operation. Even if saturation takes place at both ends the time at which it occurs will be different at each end, and transient unbalance will still occur but will be of shorter duration, and its significance will depend upon the length of this duration in relation to the energisation time necessary for relay operation. Even if the current transformers are identical and similarly loaded it is possible, although somewhat improbable, for transient unbalance conditions to occur. If we consider an internal fault occurring and being cleared, it is possible that the current transformers at the two ends will be left with a remanent flux in the cores which will be of opposite polarity at the two ends. If, as is possible, this remanent flux is retained and an external fault then occurs, the transient response of the current transformers can be quite different at the two ends and so produce large unbalances.

With dissimilar current transformer designs at the two ends the problem of transient instability will be more difficult. Fortunately, this case occurs most frequently on the higher voltage systems, for example, bushing current transformers at one end and post-type at the other. In such cases the current transformers are of relatively good performance and severe saturation under through-faults can be avoided. With modern protective-gear testing equipment it is possible to take the above aspects into account when assessing its performance capabilities in the laboratory.

From the above, it will be apparent that the evaluation of the maximum through-fault current on the circuit to be protected is of considerable importance. The magnitude alone does not define the severity, however, because the time constant of the fault current also controls the severity of transient saturation.

It is generally easier to satisfy the c.t. requirements for transient stability if 1 A secondaries are used. This is particularly so where the lead burdens are high and

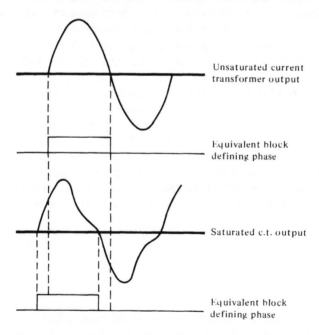

Fig. 10.9.2A   *Phase shift due to c.t. saturation*

possibly unbalanced. Since many of the schemes described operate on the phase-comparison principle at high currents, it is worth noting that this principle is still subject to the problems of transient unbalance. It is sometimes wrongly assumed that phase-comparison schemes are inherently superior to other schemes under transient saturation conditions in current transformers. The output waveform of a saturated current transformer, Fig. 10.9.2A, shows that this is not so.

If it is assumed that the phase relationship of a distorted current is defined by the points at which the secondary current crosses zero, then it is easy to appreciate that transient saturation can produce considerable phase shifts which can encroach on the setting of the protection.

10.9.3   Operating times

The operating times of the system described fall into two broad classes, 5-6 cycles for the earlier designs and 2-3 cycles for the later designs. There are two aspects concerned with operating time, first, the limitation of fault damage and, second, the limitation of disturbance to the system as a whole. For the lower voltage networks, operating times of 5-6 cycles are generally adequate to prevent damage, and system disturbance with such operating times is not significant. In fact, with cable systems, the damage caused with operating times of this order is sometimes so small as to cause difficulties in locating faults. (*Note:* these operating times are usually associated with circuit breaker tripping times of the order of 5 cycles or less.) The need for fast operating times is greater where system stability and limitation of disturbance are important, this being particularly the case with high-voltage transmission systems. With such systems, circuit breaker tripping times of 2-3 cycles are common. The need for reducing fault-clearance time is likely to continue with the trends in design of modern transmission systems because the fault levels are increasing together with the sizes of transformer and generator units. The general effect is to increase both the value of a protected unit and its functional importance, and at the same time to reduce the inherent ability of the system to withstand disturbances or to recover from them. With the present principles and techniques used for protection, it is difficult to visualise overall protection times of less than one cycle if the present high levels of protection stability and reliability are to be retained.

10.9.4   Fault settings

(*a*)   *Values:* In equipment such as generators and transformers the fault currents and the corresponding fault settings are generally related to the nominal full-load rating of the protected unit and thus to the nominal current transformer rating.

This is justified because the current under fault conditions is related to the frame size and thus to the full-load rating of the protected unit. This practice has been extended to pilot-wire protection although the fault current and thus the required fault settings are dependent on the system conditions rather than the load rating of the circuit. It is, however, important to take due account of the fact that earth-fault currents in a resistance-earthed system may be less than the full load of the circuit being protected. Therefore, even though the basic settings of the protection may be expressed in terms of rated secondary currents, it is necessary to evaluate the fault currents to be expected with minimum plant conditions on the system, providing an adequate margin to ensure positive operation.

Account should be taken of the tripping characteristics of the protection, that is, whether it is single-end tripping or double-end tripping. Under certain system conditions it is possible to get a reasonably high level of fault current infeed at one end but only a low level at the other. With a double-end tripping system or one with supplementary intertripping, the larger fault current at one end will ensure tripping at both ends. With a single-end tripping system, however, the tripping at the end associated with the lower infeed will depend on the fault current being sufficiently high at that end. If it is not sufficiently high initially, sequential tripping may occur following the tripping of the other end, this sequential tripping resulting from the redistribution of fault current. The settings normally given are inherent and may be modified in practice by factors which are discussed below.

(*b*)   *Current transformers:* Mention of current transformers has already been made. The normal errors expressed by the current transformer class can be estimated, but are generally small unless very sensitive earth-fault settings are used with high VA equipment and with relatively poor current transformers.

(*c*)   *The pilot circuit:* According to the type of system, the actual characteristics and length of the pilot can influence the effective primary fault settings. The assigned range of application is usually restricted to pilot lengths and characteristics which produce acceptable primary fault settings. In systems which use padding to a standard pilot-loop impedance it is really the pilot capacitance which modifies the settings. Schemes which do not use padding to constant loop-impedance tend to be affected more noticeably.

(*d*)   *Load bias:* With biased relays the effects of through-load currents may affect the effective fault setting. In interconnected primary systems, these effects may be neglected for phase-faults and also for earth-faults on multiple solidly earthed systems as a first approximation. In these cases, if load current is flowing, a fault fed from both ends of the feeder is generally assured on internal faults.

The difficult condition arises when an internal earth fault is of a restricted nature and such that it does not substantially modify the flow of load current. Such a condition can easily arise in a resistance-earthed system. With a pure differential system with no bias and with linear output characteristics, there is theoretically no change of fault setting due to through load. In practice some increase may be expected, particularly with high-impedance relays, as the relay current is not strictly proportional to the fault current due to the practical characteristics of the current transformers. With biased relays, still with linear output characteristics, the setting of the relay, and thus the effective fault setting is dependent on the value of through-load, the bias characteristic of the relay and, in most cases, the relative phase relationships between the load current and the fault current. A simple single-phase example is shown in Fig. 10.9.4A. The change in fault setting is shown in the polar diagram for various values of through load and phase relationships. It can be seen that the maximum setting occurs when the fault and load currents are in phase (along the real axis of the diagram) and this is readily calculable. This simple case can be extended to three-phase systems by using the equivalent single-phase ampere-turns produced on the summation transformer by the three-phase load. The relative phase-angle will vary according to the power-factor of the load and the fault current, and the particular phase which is faulted. The maximum value of fault setting, with the fault current and the load current in phase, is usually taken. As the load-bias effect and the three-phase setting are both related to the equivalent

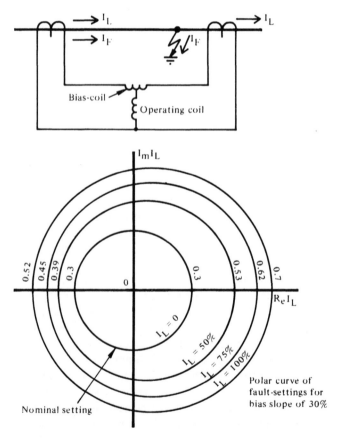

Fig. 10.9.4A   *Effect of through-load current on the primary fault setting of a linear based-differential systems*

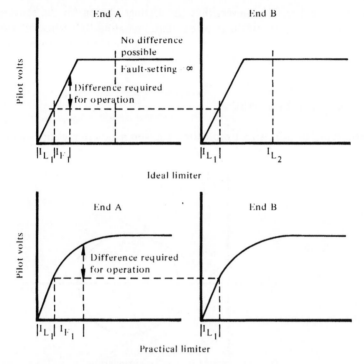

Fig. 10.9.4B *Effects of voltage limiters*

output of the summation network or transformer it is easily seen that the load-bias effect on the earth-fault setting is directly dependent on the three-phase output and thus on the fault setting chosen for three-phase faults. The proportional increase of earth-fault setting will be independent of the earth-fault setting chosen.

(*e*)   *Nonlinear limiting:* In biased systems using limiters, further modifications to earth-fault settings due to load currents are possible because of the nonlinearity of the derived signals. The worst condition is when the equivalent load ampere-turns are in phase with the fault ampere-turns. If the limiter is perfect and the combination of load and fault ampere-turns is below the limiting value the effects will be as in a linear system. If the load ampere-turns are sufficient to produce complete limiting the protection would fail to operate, as shown in Fig. 10.9.4B. As most limiters are imperfect, some increase of fault setting will occur with increasing load as shown. Care is needed therefore in choosing limiting values in relation to three-phase fault settings so that they are compatible with an acceptable increase in the effective earth-fault setting with load. The case in which load ampere-turns and fault ampere-turns are in antiphase (or in which a fault is fed from both ends) produces lower settings than the case described above and may be disregarded.

(*f*)   *Starting relays:* The above considerations apply generally to the settings obtained from the discriminating link. Where starting relays are used their settings are not so seriously affected by load currents. Such effects are usually readily predictable. It is necessary, however, that the settings chosen for the starting relays should always exceed the settings of the discriminating link when all the above factors have been taken into account. This means that the ultimate settings of the protection are always controlled by the starting relay.

10.9.5   Protection characteristics

Two basic types of differential protection characteristic are met in practice; these

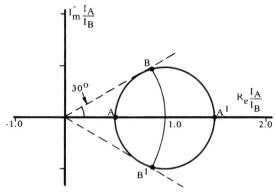

(a) Polar characteristic. (Phase comparison
characteristic shown dotted)

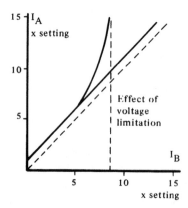

(b) Bias characteristic showing
variation of points AA^1 over
range of currents.

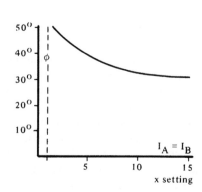

(c) Phase angle characteristic
showing variation of points
BB^1 over range of currents.

Fig. 10.9.5A   *Bias characteristics of pilot-wire differential protection systems*

being normally determined by the method in which end to end communication is achieved.

Idealised characteristics are shown in Fig. 10.9.5A (*a*) which illustrates the boundary of the two types of characteristic plotted on the complex current plane (having real and imaginary axes) in terms of the effective ratio of outputs I_A/I_B from the summation devices at the two ends of the protected line. The circular characteristic describes comparison in both phase and amplitude of the summated currents and the straight line characteristic denotes that the protection is only able to respond to a phase-angle difference between the currents at the two ends. Practical pilot-wire protection systems may have either of these characteristics, or, frequently, a composite characteristic depending upon the operating level of the protection, whereas carrier current systems, because amplitude information is not preserved in the relaying signal quantities, invariably possess the phase-comparison characteristic.

Three alternative methods of presenting the characteristics of pilot wire protection systems are shown in Fig. 10.9.5A and these comprise:

(*a*)   Polar characteristic showing the ratio of I_A/I_B defining the stability boundary of the protection.

(*b*)   Bias characteristic showing the value of I_B to produce tripping, for different values of I_A where I_A and I_B are in phase.

(*c*)   Phase-angle characteristic, showing the value of phase angle which provides tripping for a range of values of $I_A = I_B$.

The complete range of characteristics specifies the performance of the protection. Characteristic

(*a*)   illustrates the basic polar characteristic of the protection and usually applies for low levels of energisation

(*b*)   illustrates the way in which the real axis points AA'vary with operating level, and

(*c*)   shows the variation of the points BB' over the range of operating levels. Characteristics (*b*) and (*c*) thus describe the effect of voltage limiting on the protection stability zone.

Characteristics of this type provide basic design and application data.

Linear comparison over privately owned pilot circuits gives a characteristic for the comparator and pilot-wire taken together, of the kind shown in Fig. 10.9.5B (*a*) when plotted in the I_A/I_B plane.

Where pilot wire protection has to operate over a pilot circuit having an imposed voltage limitation, it is almost inevitable that phase-comparison characteristics will be obtained over the higher current levels of the working range because of the loss of amplitude information in the derived relaying signal quantities. The effect of voltage limiting on protection characteristics is shown in Fig. 10.9.5B(*b*).

Practical schemes, therefore, even when designed for maximum performance, will possess a characteristic which varies from the amplitude comparison circle to the phase comparison characteristic (Fig. 10.9.5B) when used with rented telephone type circuits.

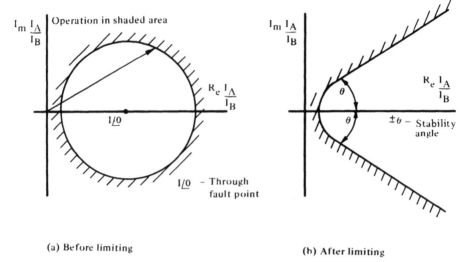

(a) Before limiting

(b) After limiting

Fig. 10.9.5B   *Complex current plane characteristics*

10.10   Power-line carrier phase-comparison protection

10.10.1   Introduction

The use of pilot-wire protection on overhead lines has already been discussed. In particular the extension of the use of pilot-wire protection for overhead lines on transmission systems at voltages above 100 kV using rented pilots has been described. The practical upper limit of about 50 km for such applications leaves the need for other forms of protection for greater distances, for example, 150 km and above. This need can be satisfied by:

(a)   The use of non- unit protection, especially of the distance class.

(b)   Unit systems employing a power-line carrier circuit as the information channel.

(c)   As in (b) but using radio link.

(d)   The combination of distance protection with a supplementary power-line carrier circuit to provide the qualities of a unit system of protection.

Chapter 7 dealt with the basic characteristics of power-line carrier transmission and the various coupling and terminal equipments. Although described in relation to digital signalling transmission, the methods and equipment are generally applicable to phase comparison protection using block modular carrier channels.

Chapter 9 covers the use of communication links, including power-line carrier, to achieve schemes of distance protection which have precise selectivity of fault-position.

This section deals with protection systems in category (b). Carrier-current equipment is rather complex and expensive and so is justifiable only on important transmission lines. In Britain the use of power-line carrier dates from the mid-thirties when the first stages of the 132 kV Grid were being formed and when the problems of the early forms of distance protection were considerable. These early applications were of the directional comparison type. From the early 1950s, renewed development of carrier-current protection to supply the needs of the rapidly expanding transmission network at 132 kV led to new systems based on phase comparison of current. This form of carrier protection was preferred to the then existing types of distance protection. The subsequent development and improvement of distance protection, together with some of the problems experienced with carrier protection, led to renewed application of distance protection, particularly on the 275 kV system. A period of application of both types of protection became desirable. Some of the limitations of the earlier generations of power-line carrier equipment have been minimised or eliminated by the more modern applications of semiconductors or integrated circuits and phase-comparison carrier systems of this kind are widely used on the 400 kV supergrid system.

All applications of power-line carrier to protection signalling take the form of a high-frequency signal, the band of frequencies employed being 70-500 kHz in the UK. The frequency bands employed abroad are generally of this same order but may differ to some extent from country to country depending on local regulations. The carrier signal is injected into the overhead line conductors at one end by means of suitable high-frequency coupling equipment and extracted by similar equipment at the other end. The information to be conveyed is superimposed on the basic carrier signal in a number of ways according to the type of protection involved. The carrier signal may exist continuously on the line or be injected for a short duration for fault conditions. If continuous, the power level of the injected signal will be low, say 1 or 2 W, or if injected for a short time will be higher, say 10 to 20 W. These are restrictions imposed to meet the requirements of the authorities in regard to other services, for example, radio, navigational aids, etc. The equipment for coupling the h.f. signal to the line must also include means for suppressing interference between the carrier on the line concerned and similar equipment on the remainder of the system. The coupling equipment is very similar for all the forms of protection using carrier, as described in detail in Chapter 7.

The use of power-line carrier in Britain has been primarily on the basis of protection requirements, the main functions of control, supervision and communication, being effected, with certain exceptions, through the telecommunication system. In many countries the main application of power-line carrier has been for the purposes of control, supervision and communication, with the protection requirements as an additional use. These differences affect the extent to which carrier is used and the economics involved, as some equipment such as that required for coupling and power supplies are common to both applications.

10.10.2 Types of information transmitted

It is not generally practical to superimpose information about the magnitude of primary currents on carrier signals although there have been a limited number of such applications abroad for very important transmission circuits. This leaves the following forms of information which can be readily conveyed by carrier signals, the systems of protection based on these being described in more detail in later sections.

(*a*)   *Information about the phase-angle of the primary current:* Information about the phase angle of the primary currents is superimposed by the block modulation of the carrier signal by the 50 Hz current concerned. It can be seen from the schematic arrangement Fig. 10.10.2A that this can be used to give a phase-comparison system of protection because the combination of the two signals is available to each end and, if processed through suitable circuits, there will be an output at each end which is dependent upon the relative phase difference of the two carrier blocks and thus of the two primary currents. The responding blocks and thus of the two primary currents. The responding relay will decide the critical angle at which tripping takes place. Schemes of this type use bursts of high carrier power, for example 20 W, during fault conditions.

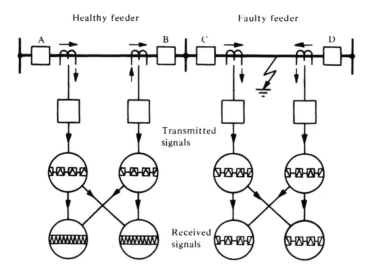

Fig. 10.10.2A   *Principle of phase-comparison protection*

(*b*)   *Information on power direction:* The direction of primary current can be compared with its associated voltage by means of a directional relay, and the information relating to the operation or nonoperation of this relay can then be sent to the remote end. The arrangement requires only an 'on-or-off' form of information to be sent through the carrier channel. This application can be considered as using the presence of a carrier signal to secure stability on through faults. The information transmitted is not quantitative, as in the case of phase-comparison, but simply indicates that the direction of the current is either 'into' the line or 'out' of it, as defined by the characteristic operating or stabilising angular zone of the directional relays. A development of this arrangement is the use of more sophisticated relays, such as distance relays, to provide the information on the direction of current.

(c)   *Intertripping signals:* Both the previous applications involved reliable transmission of carrier signals to secure stability on through faults. Such uses are of the type where the carrier is an essential feature of the protection.

The carrier signal can be used, however, to transmit positive tripping commands from one end of a line to the other. This use of carrier can be considered as supplementary, as the stability of the protection is not impaired by loss of the carrier channel. The tripping command may be arranged to take cognizance of the operation or nonoperation of some protective relay at the receiving end, this being termed 'permissive intertripping'. In other cases the tripping command is directly effective and is known as 'intertripping'.

10.10.3   Basic principles of phase-comparison protection

Information about the phase angle of the primary current is transmitted to the remote end over the carrier link by means of 100% modulation of the carrier signal, the carrier being transmitted in blocks corresponding to alternate half-cycles of 50 Hz current. Under through-fault conditions, the fault currents at the two ends of the protected line are equal in amplitude but $180°$ out of phase. (*Note:* 'in-phase' or 'out-of-phase', the instantaneous direction of current flow at an end, i.e. whether it flows from line to busbar or vice-versa). For through-fault conditions, the blocks of carrier will occur on different half-cycles at each end and the resultant carrier signal on the line will be substantially continuous. This condition is taken as the criterion for stabilising. For internal faults, the currents at the two ends of the line are approximately in phase. Blocks of carrier will occur at each end more or less at the same time and the resultant carrier signal comprises coincident blocks, i.e. the signal is not continuous. This condition is taken as the criterion for tripping. With this simple principle in mind, the various basic elements required in practice will now be considered.

The complete equipment at one end of the line can be divided into a number of functional elements as shown in Fig. 10.10.3A. The three secondary currents from the main current transformers are fed into a sequence network and starting unit.

The function of the sequence network is to extract the sequence components of the line currents to be used by the starting unit and to produce a single-phase signal which conveys the phase information of the 3-phase line currents to be used for the modulation.

The modulator is also fed with an h.f. carrier signal from the oscillator and the result of combining these two signals is to produce a 100% amplitude modulated carrier-signal. This signal is a relatively low power level and the next stage is to amplify this in the power amplifier so that a signal of about 10 to 20 W is fed to the coupling equipment and thus via the line to the remote end.

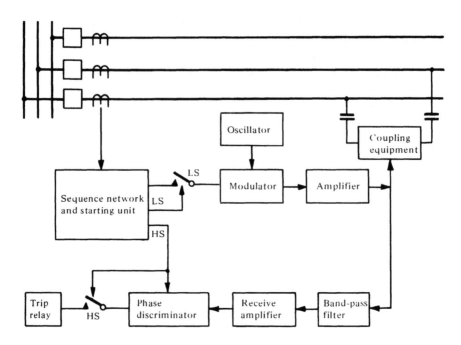

Fig. 10.10.3A   *Block diagram of phase-comparison scheme*

The carrier signal is received via coupling equipment and separated from other carrier signals by means of a narrow bandpass filter after which it is amplified in the receive amplifier and then applied to the phase discriminator. It should be noted that the received signal at this stage will contain both the remote signal and the local signal so that the phase discriminator assesses the relative phase angle of the currents at the two ends of the line by looking at the total carrier signal which exists on the line. The output of this phase discriminator is fed to the tripping relay, this output being theoretically zero if the protected circuit is healthy.

The complete equipment is controlled by means of a starting unit which feeds low-set and high-set relays. These relays ensure that carrier is not transmitted continuously, that it is transmitted during fault conditions, that both ends are always in operation at the same time, and that the fault settings of the protection give correct phase comparison when the line capacitance current is taken into account.

10.10.4   Summation networks

(*a*)   Summation techniques are involved both in the 50 Hz modulation of the carrier signal and in starting. It is necessary to ensure that the output of the summation network used for modulation is adequate for all conditions under which the starting relays operate. The obtaining of a single-phase 50 Hz signal from the three line currents requires assessment of the way in which this output is related to the various conditions of load, type of fault, the line capacitance current and combinatons of these. The following design principles are desirable.

(i)   The summation characteristics of the modulator network should be the same as those of the starting network. This ensures that if an output is obtained for modulation under a particular fault or load conditions, then a corresponding output is obtained for starting.

(ii)   The modulator network should be linear over the range of anticipated fault currents. Saturation or nonlinearity can produce incorrect phase-angle differences which can interfere with correct tripping or stabilising. These spurious phase differences may be particularly pronounced on internal faults where the fault currents at the two ends may be appreciably different in amplitude, thus producing different degrees of nonlinearity at the two ends.

(iii)   The transient response of the network should be such that the output signal is related to the 50 Hz fundamental component of the current and that the d.c. component and higher harmonics are not reproduced.

(iv)   The network should have proportionally lower outputs for load and capacitance currents in order to reduce the complicating effects of these.
In practice, it is not possible to satisfy all the above requirements and different designs of equipment may show relaxations of some or all of these.

(b)   *Summation transformers:* The simplest type of network is one using a tapped summation-transformer as described in Section 10.6.3. The outputs obtained from such a device with different fault conditions were described together with the possibility of blind spots occurring under certain fault conditions. When carrier protection is used on systems with multiple solidly earthed neutrals the possibility of blind spots is not great. It should be noted that, with such systems, the values of fault current for various types of fault are comparable with each other and the characteristics of summation transformers are not particularly suited to this requirement. Nevertheless some designs of carrier protection have successfully used summation transformers for modulation signals although, for some cases, it has been necessary to select the tapping values carefully.

(c)   *Phase-sequence current networks:* Simple examples of the principles involved in sequence networks have been described in Section 10.6.4, where it was shown that a network could be designed to give a single-phase output related only to a particular sequence component of the three-phase currents. Where a sequence network is used to provide an output for modulating carrier the modulation must be effective for all types of fault. The problem is to design a network which will be satisfactory for all or most of these fault conditions. Unbalanced faults always contain a negative-sequence component and this is the preferred output for a modulating network. Balanced faults are entirely positive sequence and the modulator network should therefore also have a proportion of positive-sequence output. The modulator network should therefore be a mixed positive- and negative-sequence type and it remains to decide the relative sensitivity to these sequence components. First, we should remember that the negative sequence component of an unbalanced fault is substantially less than the total fault current, e.g. 33% in the case of earth faults, whereas the positive sequence component of a balanced fault is 100%. Since, however, the load and capacitance currents can appear predominantly as positive sequence currents, it is desirable that the positive-sequence sensitivity should not be too high.

A mixture of sequence components which has been shown theoretically to provide satisfactory results and has given satisfactory performance in service is given by

$$Im = KI_2 - I_1$$

Where Im is the required single-phase modulating quantity, I_1 and I_2 are the positive- and negative-sequence components, respectively. The value of K is usually between 5 and 6 but consideration of the 'blinding' effect of the prefault load current with limited internal fault current suggests that some advantage may be gained by using a higher value of K under conditions of low fault current.

Fig. 10.10.4A shows the importance of checking the output for difficult fault conditions, such as double earth faults, for various system parameters. These system parameters are conveniently defined by the ratio of zero-sequence to positive-sequence impedance, e.g. Z_0/Z_1.

It also shows the polarity with which we mix the components is important. A perfect network would have the same output for all types of fault and for all values of Z_0/Z_1.

Many arrangements of sequence networks could be used to give the output required above. One example is shown in Fig. 10.10.4B.

In this network, the positive- and negative-sequence currents are extracted and summation amplifiers and a single active 60° phase lag circuit to summate voltage signals proportional to the three-phase line currents in accordance with the standard sequence component equations:

$$I_1 = \tfrac{1}{3}(I_R + \alpha I_Y + \alpha^2 I_B)$$

and

$$I_2 = \tfrac{1}{3}(I_R + \alpha^2 I_Y + \alpha I_B)$$

The summations are made as follows:
    The output of I.C.1 = $-V_Y$, this is summed with V_B and phase shifted by 60° ($-\alpha$) in I.C.2

So I.C.2 output = $-\alpha(-V_Y + V_B)$
This voltage is then summed wth V_R and V_B in I.C.3 to give an output corresponding to the positive-sequence current:

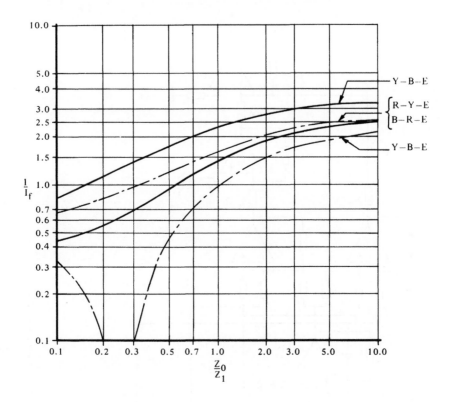

$$I = 5I_2 - I_1$$

For Y–B–E

$$\frac{I}{I_f} = \frac{6\,Z_0/Z_1 + 1}{\sqrt{3}\,\sqrt{[1 + Z_0/Z_1 + (Z_0/Z_1)^2]}}$$

For R–Y–E
B–R–E

$$\frac{I}{I_f} = \frac{\sqrt{[1 - 3Z_0/Z_1 + 21\,(Z_0/Z_1)^2]}}{\sqrt{3}\,\sqrt{[1 + Z_0/Z_1 + (Z_0/Z_1)^2]}}$$

$$I = 5I_2 + I_1$$

For R–Y–E
B–R–E

$$\frac{I}{I_f} = \frac{4\,Z_0/Z_1 - 1}{\sqrt{3}\,\sqrt{[1 + Z_0/Z_1 + (Z_0 Z_1)^2]}}$$

For Y–B–E

$$\frac{I}{I_f} = \frac{\sqrt{[1 + 3Z_0/Z_1 + 21(Z_0/Z_1)^2]}}{\sqrt{3}\,\sqrt{[1 + Z_0/Z_1 + (Z_0/Z_1)^2]}}$$

I_f = Fault current in one phase

Fig. 10.10.4A *Performance of summation networks using a combination of negative- and positive-sequence components under two-phase-earth fault conditions*

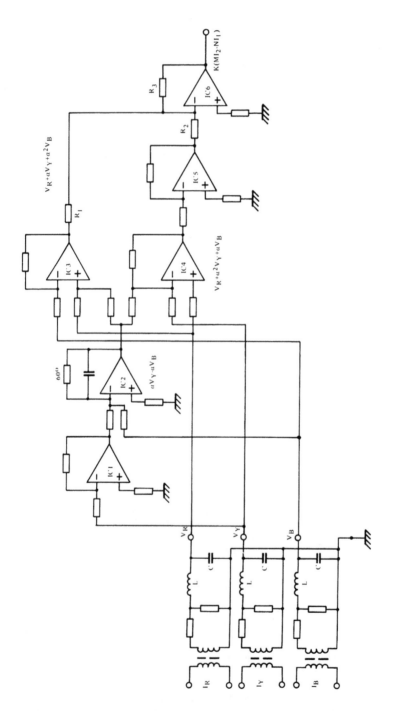

Fig. 10.10.4B   *Mixed positive- and negative-phase sequence current network used for modulation signal*

Output I.C.3 $= -\alpha\,(-V_Y + V_B) + V_R - V_B$

$\qquad = V_R + \alpha V_Y + V_B\,(-1 - \alpha)$

$\qquad = V_R + \alpha V_Y + \alpha^2 V_B = K_1 I_1$

I.C.4 produces an output corresponding to the negative sequence current by summing the output of I.C.2 with V_R and V_Y.

Output I.C.4 $= V_R - [V_Y - \alpha\,(-V_Y + V_B)]$

$\qquad = V_R + (-1 - \alpha)\,V_Y + \alpha V_B$

$\qquad = V_R + \alpha^2 V_Y + \alpha V_B = K_2 I_2$

I_2 is now inverted by I.C.5 and $K_1 I_1$ and $-K_2 I_2$ summed by I.C.6 to give the modulation signal.

I.C.6 output $= K(MI_2 - N.I._1)$

where $K = R_3, M = \dfrac{K_2}{R_2}$ and $N = \dfrac{K_1}{R_1}$.

Frequency filtering in a sequence network for modulation: Since a sequence network is a single-frequency device, it will have spurious outputs under transient and non-50 Hz conditions. In practice the transients such as the d.c. components of switching surges can produce spurious outputs and these could interfere with correct behaviour of the phase-comparison protection. It is usual therefore to include frequency filtering either at the input to the sequence network or in the modulator circuit. The sequence network shown in Fig. 10.10.4B is of the bandpass type and includes the filter at the input to the network. The input circuit consists of a transactor which provides isolation and high-pass filtering while the inductor L and capacitor C provide a low-pass characteristic resulting in an overall bandpass input filter. Such a filter should have negligible relative phase shift at 50 Hz and should have upper and lower cut-off frequencies of about 100 Hz and 25 Hz, respectively.

10.10.5   Modulation of h.f. signal

The modulator receives the 50 Hz signal from the modulator circuit via a circuit gated by the low-set starter and uses this signal to key the output from the oscillator. The resulting signal then undergoes power amplification before being fed to the line via the coupling equipment.

The oscillator is running continuously and may be one of many possible types. One of the main requirements of such an oscillator is that the frequency should be stable, for example to 1 part in 10^5. Stability characteristics of this order can be obtained by using a crystal controlled oscillator or a magnetostrictive oscillator.

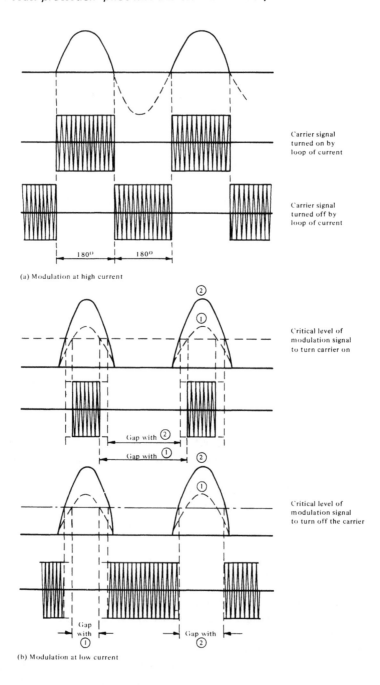

(a) Modulation at high current

Carrier signal turned on by loop of current

Carrier signal turned off by loop of current

Critical level of modulation signal to turn carrier on

Critical level of modulation signal to turn off the carrier

(b) Modulation at low current

Fig. 10.10.5A   *Effect of current magnitude on modulation*

Modern equipments make use of digital frequency division in order to allow a single crystal oscillator to be used to produce a range of carrier frequencies. There are other requirements that the oscillator must meet such as low distortion and amplitude stability but these must be considered in conjunction with the characteristics of the power amplifier which will also affect these characteristics.

It is necessary to consider how the blocks of carrier and the gaps between them will vary over the very wide range of current inputs that are met in practice. One of the requirements of a modulating network is linearity of output up to the required maximum current. If this requirement is satisfied modulation will be correct (i.e. 180° blocks) at high currents and it is only the low current end which requires examination. The process of modulation is theoretically independent of amplitude but in practice amplitude must always be taken into account at the low current end. There will generally be a critical current level at which the modulation device will key the carrier 'on' and 'off'. The modulation process usually involves taking one half-cycle of 50 Hz signal and using this to create blocks of carrier. This gives two possibilities, as shown in Fig. 10.10.5A, because the half-cycle of current can produce the requisite blocks of carrier either by turning the oscillator output on or by turning it off.

At high currents, both methods would produce the same result but their characteristics are entirely different at the low current end. As shown in Fig. 10.10.5A, if we assume a perfect keying circuit which switches instantaneously at a critical current level, the method of using the current loop to switch on the oscillator gives a carrier signal in which the gaps tend to increase as the input is reduced towards the critical level. In the other case the carrier tends, at low current levels, to a continuous carrier signal. The second method is preferred in practice because it is important that the system tends towards stability (as would be obtained with a continuous carrier transmission) at marginal conditions. The first method tends to a tripping condition.

The choice of critical current level in relation to the primary fault condition is important. In earlier schemes, this level was chosen such that the modulation was substantially complete, i.e. 180° blocks, at the low-set setting. It will be shown later how this created problems and modern systems now choose a critical level to suit the fault settings of the protection and the capacitance current of the protected circuit.

10.10.6   Junction between transmitted and received signals

With the use of common coupling equipment for transmission and reception there will be a point at which both signals meet. This point will present attenuation to both signals. Any attenuation of the received signal will not be important because incoming noise will be similarly affected and the loss of received signal strength can be recovered in the receiver gain without reducing the signal/noise ratio. In the case of the transmitted signal, the attenuation is more important because it will reduce

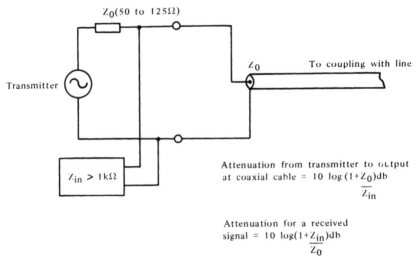

Fig. 10.10.6A   *Simple h.f. junction between transmitted and received signals*

the power to the line and will thus decrease the signal/noise ratio of the received signal at the remote end. The amount of local signal fed back to the receiver need only be small and this signal can therefore tolerate considerable attenuation at this common point.

These collective requirements can be satisfied in a number of ways, the simplest and one most used in practice is to connect the receiver directly to the output of the transmitter before being fed to the coupling equipment as shown in Fig. 10.10.6A. In this arrangement the input impedance of the receiver is designed to be much greater than the characteristic impedance of the system so that the output power lost to the receiver is very low.

10.10.7   Receiver

The design of this component will vary depending upon the particular manufacturer. In general however, it comprises a wide-band amplifier covering the whole frequency range, i.e. 70-500 kHz because it is fed with a combined signal through the requisite narrow-band filter to suit the particular installation. It must, of course, have an adequate overall gain to cover all possible practical requirements. It is normal to provide a variable attenuator at the input to the receiver so that the overall gain can be controlled to suit the needs of a particular installation. Adjustment of this attenuator is normally part of the commissioning routine to set it for the maximum sensitivity compatible with the noise levels to be expected.

The amplification referred to above is at the carrier frequency. Following amplification the carrier signal is applied to a detector or demodulator circuit, as shown in Fig. 10.10.7A, which removes the h.f. signal and leaves the 50 Hz

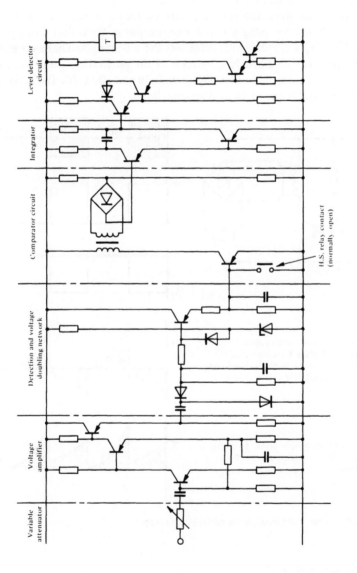

Fig. 10.10.7A   *Basic circuit of receiver*

information in the form of blocks corresponding to carrier and gaps in carrier. After demodulation, the signal is passed to a circuit which responds to the mark/space ratio of the blocks which is a measure of the phase angle difference between the currents at the two ends of the line. There are many different circuits used to respond to the mark/space ratio but they all have the common feature that an ouput is produced when the gap in carrier exceeds a certain critical level. Two typical methods are by using a transformer output in the collector of a transistor circuit or by using a circuit which integrates the waveform and has a critical detection level equivalent to the stabilising angle of the protection. These two methods are shown in diagrammatic form in Figs. 10.10.7B and 10.10.7C. The action of this stage of the receiver is one of phase discrimination.

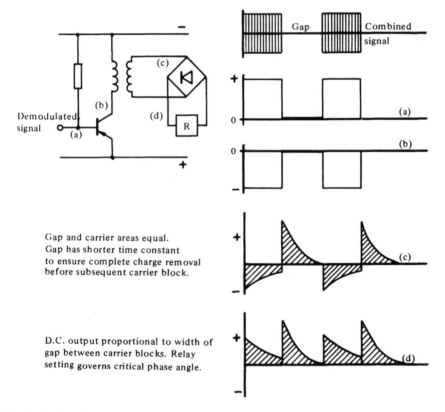

Gap and carrier areas equal.
Gap has shorter time constant
to ensure complete charge removal
before subsequent carrier block.

D.C. output proportional to width of
gap between carrier blocks. Relay
setting governs critical phase angle.

Fig. 10.10.7B   *Transformer output as phase discriminator*

10.10.8   Tripping circuit

The output of the phase-discrimination stage of the receiver will generally vary as shown in Fig. 10.10.8A, which also shows the corresponding tripping and stabilising

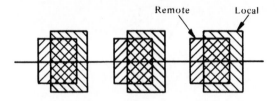

(a) Total carrier on line

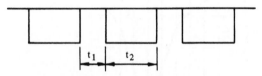

(b) Demodulated and limited signal

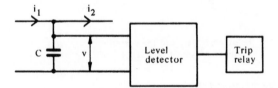

(c) Integrator as current fed capacitor. i_1 is constant charge during t_1, and i_2 is discharge current. Relative values of i_1 and i_2 decide critical angle a at which tripping occurs.

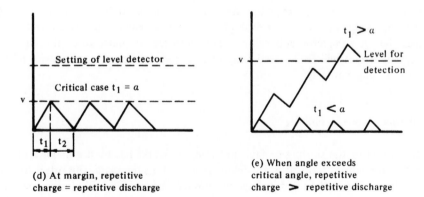

(d) At margin, repetitive charge = repetitive discharge

(e) When angle exceeds critical angle, repetitive charge $>$ repetitive discharge

Fig. 10.10.7C   *Integration and level detector*

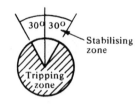

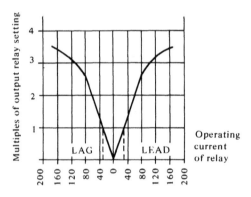

Fig. 10.10.8A   *Relay setting and stability angle*

zones of the phase angle between the currents at the two ends.

The phase discrimination is monitored at each end by the high-set starting circuit so that no tripping can take place unless the high-set circuit is operated. The setting of the tripping circuit determines the angle over which the tripping zone will extend. The choice of tripping angle will be discussed later. The form of the tripping circuit may vary, for example it may be an electronic bistable circuit or it may be an integrator and critical level detector. In most cases the electronic tripping circuit will feed a reed relay or electromechanical relay to provide a heavy duty tripping contact.

10.10.9   Starting circuits

The design of the starting or fault detecting networks has been a major feature of phase-comparison protection for many years and, in the main, this arises from the fact that most modern high-voltage systems can vary considerably in the conditions producing maximum and minimum fault currents such that the maximum load current can be of the same order as the minimum fault current.

We must obviously cater for these minimum fault currents but at the same time we must avoid continuous carrier transmission by the use of suitable starting relay settings. This is not difficult in the case of unbalanced faults but it is extremely difficult in the case of three-phase faults because these are generally indistinguishable from load currents. This has led to the use of what is known as 'impulse' or 'load-dependent' starting techniques. The basic principle used is to create a short duration condition which is dependent upon a sudden change of current magnitude rather than current magnitude itself. Fault conditions, including three-phase, will always produce this sudden change of current. Changing load-currents may also produce it but, since the starting is not continuous, the short duration of transmission is acceptable and stabilising will occur for the load conditions. Various techniques have been used to give these characteristics and perhaps if we consider the details of the two systems used in Britain the detailed working may be understood.

10.10.10   Telephase T3

The fault detection and modulation units of this system are shown in block diagram form in Fig. 10.10.10A. The line current transformers are connected to bandpass filter circuits composed entirely of passive components which produce a three-phase voltage signal which is applied to the sequence component filter. The sequence component filter realises the standard sequence component equations by means of solid-state summation amplifiers to produce output voltages proportional to the positive, negative-and zero-sequence components of the input current. The positive-and negative-sequence components are rectified by active a.c./d.c. convertors, the outputs of which are fed to low-set and high-set level detectors. In the case of the positive-sequence component the output from the a.c/d.c. convertor is always controlled by low-set and high-set impulse circuits before being applied to

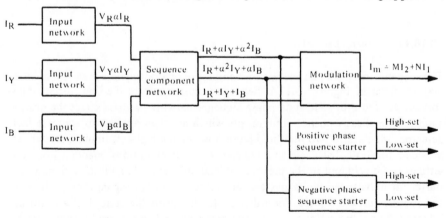

Fig. 10.10.10A   *Block diagram of phase-sequence current network for Telephase T3 protection*

the level detectors. This feature makes possible a three-phase fault setting less than full load current while avoiding continuous transmission of carrier under steady state load conditions less than full load. An impulse circuit can also be included in the negative sequence path to permit low unbalanced fault settings in the presence of considerable load unbalance.

The setting of the starting relays is controlled by adjustment of a digital thumbwheel switch which selects the proportion of the voltage from the input circuit to be applied to the level detectors. The setting of the high-set and low-set relays are kept in a fixed ratio of $1:1\cdot5$, and the fault settings provided are given in Table 10.10.10A.

Table 10.10.10A: Fault settings for Telephase T3

		Starter setting % of rated current
Unbalanced faults —	Phase-to-phase	20, 27, 34, 41, 48
impulse or nonimpulse	Phase-to-earth	36, 48, 60, 72, 84
Balanced faults — impulse	Three-phase	30, 40, 50, 60, 70

The voltages proportional to the positive-, negative- and zero-sequence components are fed into a modulation amplifier which produces a modulation signal of the general form;

$$Im = K_0 I_0 + K_1 I_1 + K_2 I_2$$

The constants K_0, K_1 and K_2 are normally preset and, as explained earlier, an acceptable modulation signal is obtained with $K_0 = 0$, $K_1 = -1$ and $K_2 = 5$. However, the constants may be altered for special applications.

10.10.11   Contraphase P10

The sequence network and starting unit employed on Contraphase is shown in block diagram form in Fig. 10.10.11A. Secondary currents from the line current transformers are fed into solid-state phase-sequence networks which extract the positive and negative-phase-sequence components which are then applied to separate fault detectors which have two threshold levels, low-set and high-set. Impulse starters are provided for both balanced and unbalanced fault conditions enabling low fault settings to be achieved in the presence of load unbalance. In addition a nonimpulse starter for unbalanced faults may also be incorporated. The settings for the high-set and low-set threshold levels for balanced faults are in the ratio $1:2$ and for unbalanced faults the ratio is $1:1\cdot5$. The balanced and unbalanced impulse starter settings, unbalanced nonimpulse starter setting and modulator threshold values are

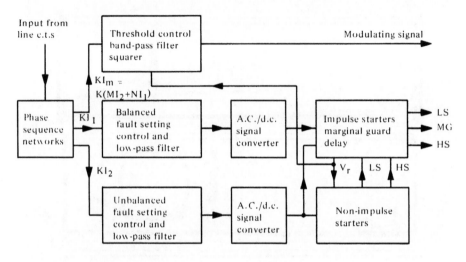

I₁ — Positive sequence current
I₂ — Negative sequence current
Iₘ — Modulating current
Vᵣ — Reference voltage

Fig. 10.10.11A   *Phase-sequence network and starting unit for Contraphase P10 protection*

set independently and the setting ranges are given in Table 10.10.11A. The equipment is illustrated in Fig. 10.10.11B.

Table 10.10.11A: Settings for Contraphase P10

		Starter setting % of rated current
Impulse starter	Positive sequence I_1	30, 35, 40, 45, 50, 55, 60
	Negative sequence I_2	5, 7.5, 10, 12.5, 15, 17.5, 20
Nonimpulse starter	Multiple of impulse starter I_2	1.5, 2, 2.5, 3, 4, 5.8, 8
Modulator network		7.5, 8.75, 10, 11.25, 12.5, 13.75, 15

10.10.12   Marginal guard

Both P10 and T3 protections also employ a technique known as marginal guard.

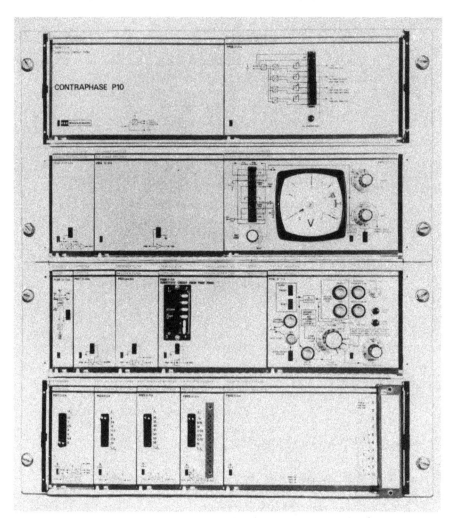

Fig. 10.10.11B   *P10 phase-comparison protection (GEC Measurements Ltd.)*

The purpose of it is to ensure the stability of the protection after the clearance of an external fault.

When an external fault is cleared, the starting circuits of the protection will start to reset; however, due to differences between the equipments and possible differences in current at the two ends (only one end may be supplying the charging current), the starting relays of one equipment may reset before those at the other end. Clearly, if both low-set and high-set relays reset at one end before high-set relay at the other, the slow end will operate. To guard against this the resetting of the low set relay is arranged to transmit a continuous carrier signal for a set period (180 ms for T3, 100 ms for P10), thus ensuring that, if a high-set relay is still operated, no protection operation will occur. This period of continuous carrier transmission is termed marginal guard. However, the inclusion of marginal guard in these protection systems can, under certain circumstances, cause slow clearance (greater than marginal guard time) of internal faults at one end. To ensure this does not happen, the marginal guard is muted when the protection operates, thus ensuring that the other end, if it has not already tripped, is not prevented from so doing by the receipt of a marginal guard signal.

10.10.13 Checking and testing

Carrier-current protection uses equipment which is complex and which comprises many components and interconnections. It has been general practice, therefore, to provide extensive facilities to check and test the soundness of the equipment, including transmission of carrier over the line itself. Although the details of such facilities vary with individual designs, they can be classed as follows:

(a) links, test switches and such facilities for use during commissioning or routine maintenance

(b) special test points at critical circuit positions to allow checks for correct operating levels to be made

(c) built-in test metering to allow power supplies, transmitted and received power levels etc., to be easily measured

(d) power-supply failure alarms

(e) built-in manually operated test facilities for low-level power-frequency injection to test, for example, trip operation or phase-comparer angle setting

(f) systematic scheme testing initiated either manually or automatically via a clock signal which provides a regular check that the equipment functions correctly.

This is an important aspect of the test function and will be considered in greater detail.

Systematic checking: These schemes have the object of testing, functionally, as much of the protection as possible under normal healthy conditions on the line, that is when the starting relays are unoperated. It is a requirement of all such testing schemes that they should be capable, at any time during their cycle of operation, of being overridden by a real fault condition. Furthermore, the test sequence and test facilities are required to cater automatically for restrictions on the time of transmission of power to the line imposed by national telecommunications authorities. As the facilities are purely test facilities they cannot safely be arranged to operate starting relays or check actual tripping relay circuits. They will, however, check nearly all the electronic carrier equipment and will prove the correct transmission and reception of carrier over the overhead line. The test will also cover loss of carrier not only due to equipment failure, but also due to the conditions of excessive attenuation on the line itself. The following systematic checks are usually provided:

(i) *Reflex testing:* This test is arranged to check nearly all of the electronic circuitry and prove the transmission and reception of carrier over the line. The reflex test can be operated manually or automatically by means of a clock built into the carrier equipment. When the test is initiated a burst of carrier is trans-mitted by end 1 momentarily operating its marginal guard circuit, and the carrier is received by a test receiver at end 2. After a short interval, end 2 responds by transmitting carrier back to end 1 which has been primed to act on the received signal to indicate a Reflex Test Correct condition. Failure of the transmissions to be received at either end results in a Reflex Test Fail indication and the opera-tion of a hand-reset relay for alarm purposes.

(ii) *Stability-on-load or extended transmission reflex test:* A modification of the Reflex Test circuits provides for transmission of carrier from both ends at the same time; modulated in accordance with the primary current. The test, which can only be operated with the protection switched out of service, produces a burst of h.f. transmission from both ends simultaneously.

This facility allows the phase difference between modulated carrier transmissions from the two ends to be observed on an oscilloscope and correct polarity of current-transformer connections proved during commissioning.

(iii) *Steady carrier transmission:* This test provides for transmission of unmodulated carrier to allow measurement of the transmitted and/or received carrier levels. This facility is required particularly during commissioning.

(iv) *Injection testing and monitoring facilities:* The test methods mentioned above do not include the starter circuits hence an a.c. injection circuit is provided to inject current into the sequence network input. The test is called a Trip Check Test and the current level is sufficient to operate the high-set starter and conse-quently allow operation of the equipment under simulated system fault conditions.

Other facilities generally provided are as follows:
(a)   Protection out of service; this facility switches the protection out of service to provide for the out of service tests, e.g. stability on load and trip check test. Remote indication of the protection-status is provided.
(b)   Indication unit; this allows the various a.c. and d.c. voltage levels in the equipment to be measured, including the h.f. carrier level.

10.11   Problems of application of phase-comparison feeder protection

10.11.1   General

Phase-comparison protection, like many other forms of protection, has some problems in its application. The initial choice depends upon many factors including economic considerations, the general policy of protective gear practice on the network and the other possible uses of carrier on the system such as communication and telemetering. It is not proposed to go into this here but there are some problems of a technical nature which affect the use of phase comparison protection and it is advisable that these be understood. The more important of these are dealt with in the following Sections.

10.11.2   Attenuation over the line length

As previously seen, attenuation along the line is important because it governs the signal/noise ratio in the receiver. Attenuation is a function of the carrier frequency, the line configuration and the weather conditions. Using phase-to-phase transmission and allowing a typical attenuation value, the maximum length of line which could normally be protected is of the order of 180 km, using power levels of about 10 W on the line. It is important to appreciate that to extend this length by increasing transmitter power is only partially effective even if regulations do not restrict such power. For example, to double the line length and maintain the same signal strength at the receiver we would have to increase the power by 10 times. Wherever possible, the lower frequency allocations should be reserved for the longer line lengths in order to reduce attenuation.

Line length also affects other aspects of phase comparison protection such as phase-angle and fault settings but these will be covered in the other Sections.

10.11.3   Tripping and stabilising angles

The choice of tripping and stabilising angles is difficult because the differences in phase angle between the currents at the two ends may vary considerably on internal faults. This is especially true if we take into account high resistance faults, extremes of plant and broken conductor or multiple earth faults. However, although it is not possible to analyse all of these conditions for all applications and arrive at specific requirements it is known that the tripping angle should be as large as possible to cater for as wide a range of conditions as possible. Thus the problem is defined in terms of the stability requirements by making allowance for all the phase differences which can occur between ends on external faults and allowing some margin on these to ensure stability. For example, a line 100 miles long would incur a propagation delay of 0·5 ms which is equivalent to 9° phase shift on a 50 Hz basis. An additional 20° is allowed to cater for current-transformer and protection equipment errors and includes a margin for safety.

10.11.4   Fault settings related to capacitance current

Capacitance current has a pronounced effect on the fault settings which may be adopted for a particular application. In general, the use of impulse starting does not alter the considerations to any appreciable extent. Basically, we must take into account two particular conditions which might occur at the level of fault settings under external fault conditions. By examining these we can give the minimum fault settings which are compatible with stability. We cannot reduce these, so it is then necessary to examine whether these fault settings are adequate for our minimum internal fault currents. It is often the case that the fault current at one end is considerably less than at the other and we may have to rely in this case upon sequential clearance of the internal fault.

The two conditions covering settings are as follows:

(*a*)   a fault condition limited by a high-resistance fault

(*b*)   a fault current limited by the source impedance under minimum plant conditions.

These conditions are shown in Fig. 10.11.4A.

In the first case, the effect will be equivalent to a phase shift between the currents at the two ends and in the second case will be equivalent to a magnitude difference. The actual numerical restrictions that these place on fault settings depend upon the type of summating circuits used for the starting relays and the modulator.

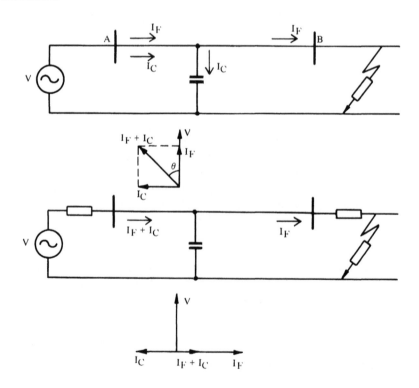

Fig. 10.11.4A   *Effect of line capacitance current*

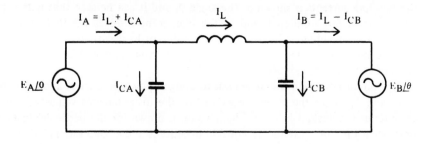

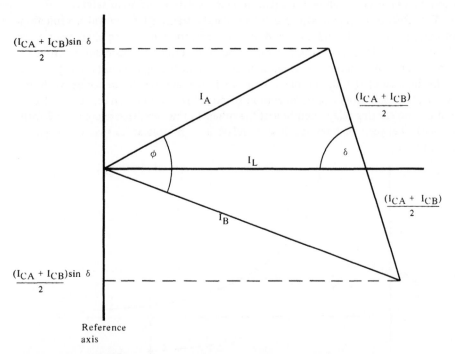

Fig. 10.11.4B *Phasor diagram showing effect of capacitance current* $(I_C = I_{CA} + I_{CB})$

The phase shift caused by capacitance current is generally catered for by the inclusion of compensation techniques.

Phase difference is most important during the conditions following clearance of an external fault when the starters have not yet reset and there may be little or no load current. The vector diagram in Fig. 10.11.4B is constructed from a general consideration of generation at both ends of the line. The phase difference between the two line currents is shown as the angle ϕ, and it can be seen that with capacitance current fed from each end of the line, ϕ increases as the load current decreases. The maximum value of ϕ, for any value of load current is obtained when half the line capacitance current is fed equally from each end, and the arbitrary angle of capacitance, δ, is $90°$.

The vector system in the diagram is rotating at power frequency and projections of I_a and I_b on to the reference axis show the instantaneous values of the line currents at any instant in time. The diagram is drawn for the particular instant in time when the line currents are equal in magnitude and of opposite polarity. It can be seen that if a bias current is applied to the modulators equal to half the capacitance current, such that the bias must be exceeded to allow modulation, as modulation occurs at one end a stabilising signal is sent from the other end. Hence, the protection will be stable for all conditions of capacitance current and through load current magnitude. In practice, the modulator bias is made equal to the capacitance current to allow for circuit tolerances with a margin of safety.

The effect of the modulator bias is to make the transmitted gap width a function of current magnitude and this modifies the protection stabilising characteristic as shown in Fig. 10.11.4C, where it can be seen the effect of capacitance current is compensated throughout the whole characteristic, including the fault setting.

In the case of (b), the problem is to avoid an amplitude difference between the two ends which would cause operation of both low-set and high-set relays at one end and no starting relay operation at the other. In this case, we are concerned with the ratio between high-set and low-set relays and the tolerances that we assign to these settings.

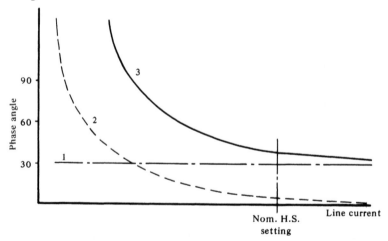

1. Displacement allowed for propagation delay and circuit tolerances.

2. Displacement due to capacitance current.

3. Operating characteristic of the protection.

Fig. 10.11.4C   *Stabilising characteristic*

The worst case of magnitude difference is obtained when the line capacitance current is in antiphase with the fault current and this condition with the starter settings and tolerances is shown in Fig. 10.11.4D.

With a per-unit value of low-set setting the high-set setting is 1·5 and with 5% tolerances the capacitance current must not exceed 0·375.

That is, the high-set setting must be greater than 1·5/0·375, equal to 4 times the line capacitance current.

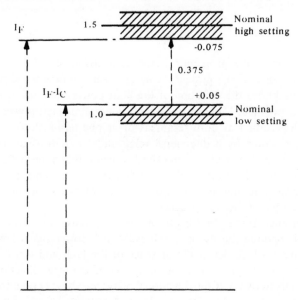

Fig. 10.11.4D *Permissible settings determined by starting relay ratios*

10.11.5   C.T. requirements

Phase-comparison protection is of the high-speed class and as such demands reasonable performance of c.t.s under transient conditions. It has been shown elsewhere that it is wrong to assume that a phase comparison feature is not susceptible to transient differences between c.t.s. The requirements will depend upon the design of equipment and application details such as lead burdens, etc. Most manufacturers provide design criteria for c.t. requirements.

10.12   Directional comparison protection

10.12.1   General

This system was the first application of carrier to protection and, although it has been mostly superseded by phase comparison and combinations of distance relays and carrier, it is still an important form of protection. It is often applied overseas, for example, in America, where it is frequently used for the earth-fault feature of line protection. The carrier signal is keyed 'on' or 'off'' according to the state of the relays at the two ends, and the information transmitted is therefore very simple. Some reference to directional comparison protection is contained in Chapter 9 because of this special relationship to some forms of distance protection.

10.12.2   Basic principles

The basic principle of the protection is the comparison of the direction of fault power at the two ends of the protected line by means of directional relays. Under through-fault conditions the direction of the fault power must be outwards at one end of the protected line. Under internal-fault conditions, fault power must be fed only into the line, either at both terminals or at one terminal. The detection of outward flowing power by a directional relay under external-fault conditions is used to initiate transmission of a carrier signal which can then be arranged to block tripping on the line both at the local end and at the remote end. The remote end, which feeds power into the line, is capable of knowing that the fault is external only by the reception of the carrier signal.

The usual arrangement is that a carrier signal is transmitted only during fault conditions. This permits the use of a relatively high-power signal level of 10 W. Relays are required which detect the presence of the fault and which prepare the tripping circuit. The tripping circuit is completed if the local directional signal is received. The characteristics of the directional relays are those chosen so that operation is obtained for power flow into the line and restrained for power flow out of the line. This arrangement is shown diagrammatically in Fig. 10.12.2A.

In the early forms of this protection the starting relays were of the overcurrent type for all kinds of fault. The earth-fault settings could be made less than full load but the phase-fault settings were of the order of 200% maximum load current or more, in order to avoid the continuous transmission of carrier under load conditions. With the development of transmission systems, the diversity of load and fault currents became such that the maximum loads and the minimum fault currents (with reduced plant) became comparable, and the above-mentioned phase-fault settings were not acceptable. As previously explained, the same conditions were a problem to the application of phase-comparison protection. In directional-comparison protection this problem is generally solved by replacing the overcurrent starting relays by distance relays which are not entirely current dependent but which have a specific impedance setting, virtually independent of load conditions. Current dependent relays can still be used for earth faults.

All the directional comparison systems use a single carrier frequency. The system is basically single-end tripping, that is there must be a flow of fault current to operate the starting relays at any given end before that end can be tripped. With interconnected systems this is acceptable.

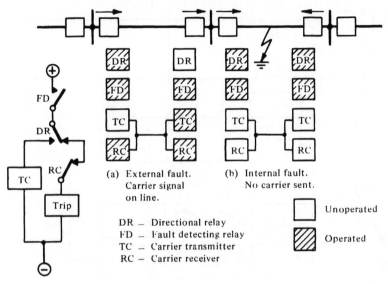

(a) External fault.
 Carrier signal
 on line.

(b) Internal fault.
 No carrier sent.

DR — Directional relay
FD — Fault detecting relay
TC — Carrier transmitter
RC — Carrier receiver

Unoperated

Operated

(c) Basic d.c. circuits

(Relay DR is in position corresponding to power flow into line. All other contacts shown in de-energised position)

Fig. 10.12.2A *Principle of directional comparison protection*

10.12.3 Basic units

A block diagram showing the principal units in a scheme is given in Fig. 10.12.3A.

Fault detecting relays comprise low-set and high-set elements. The low-set relays control a carrier transmitter made up of an oscillator and power amplifier, the carrier signal being fed into the line via coupling equipment and transmitted to the remote end. A similar carrier signal from the remote end is received through a bandpass filter, amplified in a receive amplifier and operates a carrier-receive relay. Operation of this latter relay opens the circuit of the trip relay. The high-set fault-detecting relays prepare the trip circuit which is completed upon operation of the directional relays, provided the carrier-receive relay is not operated.

A number of the above units are very similar to those used in phase-comparison protection. Only those units particular to directional comparison are dealt with in the following Sections.

10.12.4 Directional relays

The settings and characteristics of the directional relays must be chosen to cover all types of fault and fault conditions. It is generally easier to meet these requirements with a multiple-relay arrangement, for example, four relays, three for phase-faults

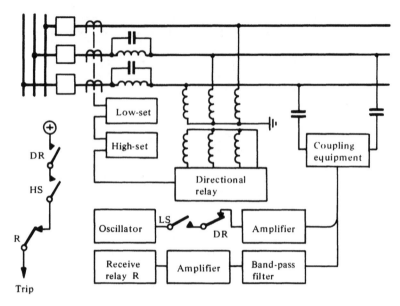

Fig. 10.12.3A   *Block diagram of directional comparison protection*

(phase-to-phase and three-phase) and one for earth-faults, rather than with only one complex polyphase relay. The characteristics chosen are generally similar to those used in directional elements of directional overcurrent and earth-fault protection except that high speed of operation is required. Various types of relay movement have been used including induction and balanced-beam types. Typical characteristics are shown in Fig. 10.12.4A.

Where distance type starting relays are used, care is necessary to match the directional and distance relay characteristics to ensure correct operation under all relevant fault conditions.

10.12.5   Fault detecting

Since the directional relays in this type of protection require voltage supplies, these same supplies are available to meet the difficult starting conditions referred to above. These conditions are fairly general in Britain and, in practice, a simple scheme based on overcurrent starting is seldom used. The combination of voltage and current in the starting relay in the form of a distance relay is now fairly standard. It has been noted that, in phase comparison, where the voltages are not necessarily available, much more complex starting equipment has been developed. It is possible to use overcurrent relays in the case of earth faults as these are easily detected by such means.

The starting relays are of the two-stage type, for example, high-set and low-set, as explained in Section 10.5. In the case of distance starting the high-set element is

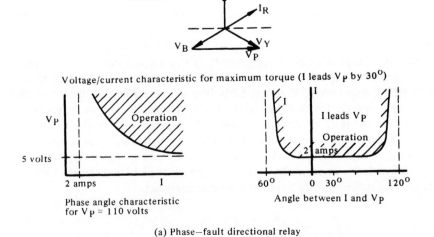

(a) Phase–fault directional relay

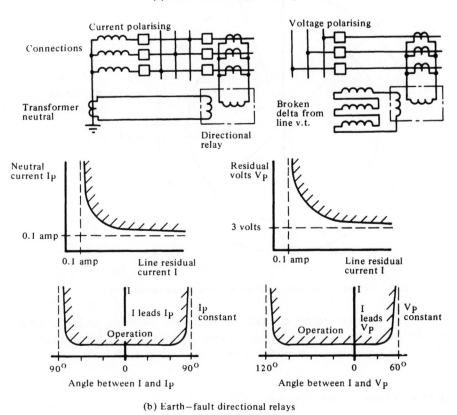

(b) Earth–fault directional relays

Fig. 10.12.4A   *Phase and earth-fault directional relays*

given a lower impedance setting than the low-set, the characteristics being as shown in Fig. 10.12.5A.

It is not, as a rule, necessary to provide a range of impedance settings on the relays to cater for the various system conditions or lengths of line. In general, one impedance setting is adequate to distinguish between load and fault conditions in all cases. As a typical example for phase-fault starting elements, the impedance settings of the starting relays would be 360 and 480 primary Ω for the high-set and low-set relays, respectively. These are fairly long equivalent line lengths so that the relays would operate for faults at fairly remote points. It must be remembered, however, that the reach of these relays would be reduced if the fault is external to the line when there may be a multiplicity of infeeds to the fault. The accuracy and characteristic range requirements of these relays are not excessive, typical values being 10% and 5%, respectively. It will be noted that plain impedance characteristics are used rather than the complex mho characteristics.

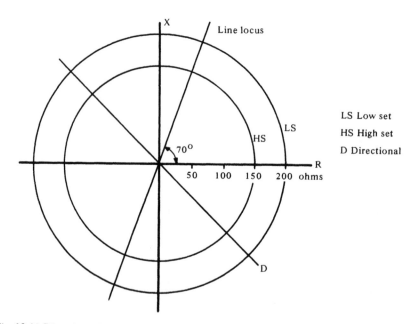

Fig. 10.12.5A   *Impedance diagram for directional comparison protection*

Where overcurrent relays are used for earth-fault detection, the adjustment provides for a range of fault settings between 20 and 100% of full-load current. The ratio between high-set and low-set relay settings is about 1·5.

10.12.6   Change of fault direction

On an interconnected system there is a possibility that the direction of through-fault power-flow in a healthy feeder may change during an external fault when a circuit breaker clears one source of supply to the fault. If the healthy line is protected by direction-comparison protection, then under these circumstances, the directional relays at the two ends of the line suddenly reverse their roles. The one that was initially restraining now operates and vice versa. If there are slight differences in timing, it is possible for the carrier stabilising signal to be momentarily interrupted and unwanted tripping to take place. To avoid this occurrence, a relay is fitted which opens the trip circuit after an external fault has been present for a short time. Fig. 10.12.6A shows this condition.

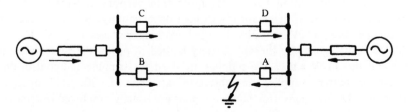

(a) Initial conditions. Directional relay at C operates.
Directional relay at D restrains.

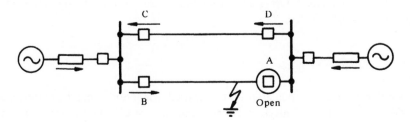

(b) Conditions following tripping of A.

Directional relay C restrains.
Directional relay D operates.

Fig. 10.12.6A   *Change of fault direction*

10.13   Power supplies

10.13.1   General

The power supplies for carrier equipment are characterised by a need for high reliability and dependability. It is essential that they be maintained in the event of failures in the auxiliary mains supply or during conditions of fault when this supply may be momentarily lost. This dependability is particularly necessary in those schemes of protection in which the carrier signal is used for stabilising purposes. In such cases it is necessary to cater for fairly long interruptions of the auxiliary supply, that is failure, as otherwise the protection would have to be taken out of service in order to avoid false tripping on through faults. In the case of impulse starting such false tripping could also occur on increasing load current.

Different practices are used to satisfy the requirements of power supplies and these differ somewhat between the United Kingdom and abroad.

10.13.2   Station battery supply

The use of the station battery for protection purposes is widespread as this supply has the merit of being already of high dependability. Battery voltages may be nominal 48V, 110V or 220V (normal working voltages of 54V, 125V and 150V), although generally the 48V battery is used to power the solid state equipment and the 110V and 220V supplies are used for tripping and control duties. The station battery supplies are subject to variations of -20, $+25\%$ of nominal voltage and d.c./d.c. convertor power supplies are usually employed to remove the effect of such variations. Station batteries also supply many inductive loads and have a large secondary wiring network associated with them. These conditions can give rise to voltage surges on the supply and the d.c./d.c. convertors include within their design adequate filtering and protection to ensure that the operation of the equipment is not affected.

10.13.3   Separate batteries

In some countries, particularly in the United Kingdom, separate batteries are provided for telecommunication service and, in some instances for protection purposes. These batteries generally have lower voltage variations and, because their use is restricted to telecommunication or protective equipment, are not subject to the same levels of interference as station batteries. Nevertheless, it is common practice to employ d.c./d.c. convertor power supplies within the equipment being supplied from such batteries.

10.14   Bibliography

Books

Carrier communication over power lines by H K Podszech (Springer-Verlag)
Teleprotection CIGRE Committee No. 34 and 35, Report of the joint Working Group on Teleprotection (March 1969)
Protective relays: their theory and practice by A R Van C Warrington (Chapman & Hall, 3rd Ed., 1977) Vol. 2, Chapter 15 *Pilot differential protection* by J Rushton

Papers

'Information links' by L M Wedepohl (*Int. J. Electr. Eng. Educ.*, 1964, 2, pp.179-196)

'The fundamental characteristics of pilot-wire differential protection systems' by J. Rushton, (*Proc. IEE*, 1961, **108**, (41), pp. 409-420)

'Fault-performance, analysis of transmission-line differential protection systems' by J Rushton, D W Lewis and W D Humpage (*ibid.*, 1966, **113**, (2), pp. 315-324)

'Review of recent changes in protection requirements for the CEGB 400 and 275 kV systems' by J C Whittaker (IEE Conf. Publ. 125, Development in power system protection, 1975)

Chapter 11

Overvoltage protection

by L. Csuros

11.1 Overvoltage phenomena in power systems

11.1.1 External overvoltages (lightning)

Overvoltages due to lightning have their cause external to the system, therefore they are often referred to as external overvoltages.

(*a*)   *Origin and mechanism of lightning:* Lightning originates from thunder-clouds which usually contain positive electric charge at the top and negative charges at the bottom. As the result of these charges, electric fields are built up within the cloud, between clouds and between the cloud and earth.

The mechanism of the lightning discharge has been studied with rotating cameras, and Fig. 11.1.1A illustrates a typical record by Schonland obtained with such a camera.[1] It has been found that the discharge is normally initiated from the cloud at a point of high electric stress, and in the majority of cases, a negative charge from the cloud proceeds towards the ground in a series of jerks of 'steps'. This is the 'leader stroke'. When the leader stroke approaches the ground, high electric stresses develop above the ground and an upward streamer of positive charges develops from the ground. (Protruding conducting objects help the development of such a streamer). When this meets the down-coming leader stroke, a conducting path is established between the thunder-cloud and earth and in this second stage a heavy discharge takes place. This is called the 'return stroke'. The phenomenon may repeat itself producing so-called 'multiple strokes'. Multiple strokes consist usually of two to three discharges but records have been obtained with as many as 40 subsequent discharges. For the power engineer, the return strokes are of interest as they carry the heavy discharge currents and produce all the manifestations associated with lightning. They are usually called simply 'lightning strokes'. The magnitude of current in a lightning stroke may vary from a few thousand amps up to perhaps 150 000A or even more in extreme cases. The average value might be of the order of 20 000A. The duration of the lightning current varies from perhaps 20-30μs up to several thousand microseconds (particularly with multiple strokes). The explosive effects are related to the magnitude of the lightning current. High current values are normally associated with short duration strokes. High energy is, however, normally associated with long duration strokes of moderate current magnitude. These have high incendiary effects and are often called 'hot lightning'.

(*b*)   *Effects of lightning strokes in power systems:* Direct lightning strokes terminating on phase conductors, earth-wires or towers of overhead lines can produce excessive serge voltages on the system. On lower voltage systems, lightning strokes near a power line (that is indirect strokes) may induce excessive voltages on the line and these surges may cause flashovers on the overhead line. Surges caused by both direct and indirect strokes may travel along the line and may cause breakdowns at the terminal equipment.

When the phase conductor of an overhead line is struck by lightning, the conductor voltage rises to a value equal to the product of the lightning current and the effective surge impedance of the conductor on which the lightning stroke terminates (see Section 11.2.1). As the surge impedance of a power line is usually around 400 or 500 Ω (the effective surge impedance is half of this value since the line sections on each side of the point where the stroke terminates are paralleled), a voltage surge of very high magnitude is initiated at the point where the line is struck. Flashover will normally occur unless the line insulation to earth is in the million volts region.

When a stroke takes place to an earth-wire or tower, the potentials along the current path may be raised to very high values. If the product of the lightning current and the effective impedance to earth at any point along the path is high enough to break down the insulation, flashover will take place either from the

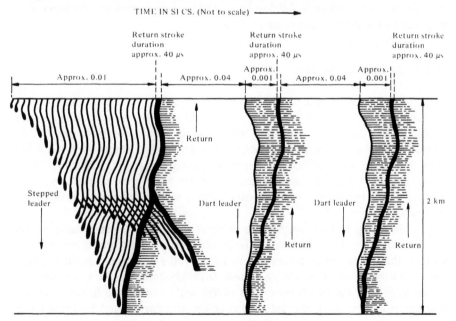

Fig. 11.1.1A   *Lightning flash to earth recorded with a rotating camera*

earth-wire or from the tower to the phase conductors (usually across the insulators). This type of lightning fault is called a back-flashover. A precise treatment of the phenomenon is very complicated, particularly if the towers are tall. It is necessary to take into account that the surge voltages and currents on the earth-wires will induce voltages on the live phase conductors, due to both capacitive and inductive coupling, and these transients will be superimposed on the power-frequency voltages.

On distribution lines operated at 33 kV or below a lightning stroke to a tower or earth wire in general causes a back-flashover. The insulation strength of lines operating at 132 kV and higher voltages is higher and the risk of back-flashovers is a function of the potential difference between the live conductors and the earthing system consisting of the towers and earth-wires. Tower footing resistance plays an important part in the considerations and, if it is high, back-flashover might take place at relatively low values of lightning current.

In respect of published information dealing with the complex phenomena occurring on overhead lines when struck by lightning or subject to induction due to nearby lightning strokes reference is also made here to the very extensive literature discussing both the parameters of lightning and their effect on overhead line performance.[1-26,72]

A flashover on an overhead line results in the tripping of the line circuit but with modern fast-acting protection permanent damage seldom occurs.

Lightning strokes to substations may cause transient overvoltages in a somewhat similar way to those to overhead lines. If insulation breakdown occurs not only may supply interruptions result but serious damage to expensive substation plant may occur with more serious consequences.

11.1.2   Internal overvoltages

Switching operations, sudden changes in system parameters, fault conditions or resonance phenomena may cause overvoltages. As these overvoltages are generated by the system itself, they are often referred to as internal overvoltages.[27-40] To provide an adequate survey of the phenomena would fill a whole book but a few cases of 'classical' types of internal overvoltages are discussed in this Section.

(*a*) *Transient internal overvoltages:* There are two types of switching overvoltages which require particular attention, both due to switching and these are discussed below.

(i)   *Switching out of transformers or shunt reactors:* The magnetising current of transformers or shunt reactors may be forcibly 'chopped' before the instananeous value of the 50 Hz current reaches a zero value. As the flux density in the magnetic core is related to the magnetising current such 'chopping' leaves some magnetic energy trapped in the transformer. This energy will be dissipated in the form of a damped oscillation since the inductance of the transformer and its capacitance, together with the capacitance of the connections to the transformer, represent an

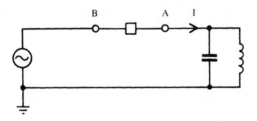

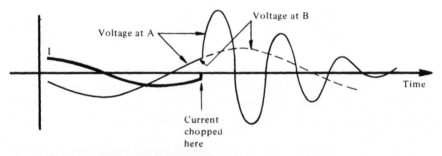

Fig. 11.1.2A *Fundamental phenomena due to chopping the magnetising current of a shunt reactor or unloaded transformer*

oscillatory circuit.[27,28,32,33] The fundamental phenomenon is shown in Fig. 11.1.2A, which is typical of the performance of air-blast circuit breakers. Oil circuit breakers seldom succeed in clearing the circuit at the first attempt at current chopping. Re-ignitions occur across the breaker contacts producing a saw-tooth voltage shape. The final chopping nevertheless occurs as shown in Fig. 11.1.2A.

During the oscillations, the energy trapped in the core of the transformer or shunt reactor appears alternately as magnetic energy in the core and electric energy charging up the capacitance. On this energy basis, the approximate magnitude of the overvoltage can be calculated in a very simple way.

The magnetic energy at the instant of chopping the magnetising current equals $i_{ch}^2 L/2$, where i_{ch} is the value of the magnetising current at the instant of chopping and L is the inductance of the transformer or shunt reactor.

The electric energy at the moment of maximum overvoltage equals $V_o^2 C/2$, where V_o is the peak value of the oscillatory overvoltage and C is the equivalent capacitance of the windings and connections to the transformer or shunt reactor.

If the electric energy at the instant of chopping is neglected (this is justified if the system voltage is small compared with the overvoltage V_o), then the peak value of the magnetic energy must be equal to the peak value of the electric energy, that is

$$\frac{i_{ch}^2 L}{2} = \frac{V_o^2 C}{2}$$

that is,

$$V_o = i_{ch} \sqrt{\frac{L}{C}} \qquad\qquad 11.1.2.1$$

The natural frequency of the oscillation is determined by the capacitance and inductance. Neglecting losses and assuming both L and C to be lumped, the natural frequency f_n of the osciallations will be

$$f_n = \frac{1}{2\pi\sqrt{LC}} \qquad\qquad 11.1.2.2$$

It can be seen from eqn. 11.1.2.1 that for a given L, that is a given transformer or reactor, and a given i_{ch} the magnitude of the transient overvoltage depends on the capacitance of the circuit. The overvoltage can be reduced by increasing the capacitance, for example by inserting a length of cable between the circuit breaker and transformer.

Nevertheless, as seen from eqn. 11.1.2.2, increasing the capacitance reduces the natural frequency of the oscillation and might increase the magnitude of the current which the circuit breaker can chop successfully. This fact is often overlooked by authors of papers on the subject. There is usually a particular value of capacitance for a given transformer or shunt reactor and a given circuit breaker which gives the maximum overvoltage. This will *not* necessarily occur at minimum capacitance.[28]

One significant aspect of this overvoltage phenomenon is that it occurs after the transformer or shunt reactor is already disconnected from the system. The live system, therefore, cannot be affected by these surges nor will a flashover or the operation of a co-ordinating gap at the transformer or reactor put an earth fault on the system (see Sections 11.4.3 and 11.5.1(iii)).

(ii)   *Switching of capacitors and unloaded feeders:* During the interruption process in a circuit breaker, the contacts separate gradually. If at any instant during this process, the voltage across the contacts exceeds their insulation strength at that instant the arc will be re-established, that is, the circuit breaker will restrike.

When a capacitor or an unloaded feeder (overhead or underground) is switched, repeated restrikes in the circuit breaker may produce overvoltages as shown in Fig. 11.1.2B. The 50 Hz system voltage is at peak value when the current passes the zero value. The circuit is easily interrupted at that moment, since, for a fraction of a cycle thereafter, the voltage across the open contacts of the circuit breaker will be relatively small. The capacitor will be left charged at peak system voltage E, while the system voltage will fall towards zero. After half a cycle, the system voltage will be $-E$ and the voltage across the circuit breaker contacts will be $2E$. If a restrike occurs at that instant, the capacitor will attempt to follow the system voltage via a damped oscillation, as shown in Fig. 11.1.2B, the peak value of which is $-2E$ relative to the system voltage or $-3E$ relative to earth. The frequency of this oscillation is determined largely by the system inductance and the capacitor or line being switched. If the arc is extinguished at the first negative voltage peak (when the oscillatory current is zero), as shown in the diagram, the capacitor remains charged at $-3E$. If the arc is extinguished at that moment and a restrike occurs again half a cycle later then the transient voltage may reach $5E$. Assuming that the arc is extinguished once more and restrikes again after another half cycle, the magnitude of the transient would reach a value of $-7E$. The dotted lines in Fig. 11.1.2B indicate the prospective values of the oscillatory transients if the arc would not extinguish at the peak voltage.

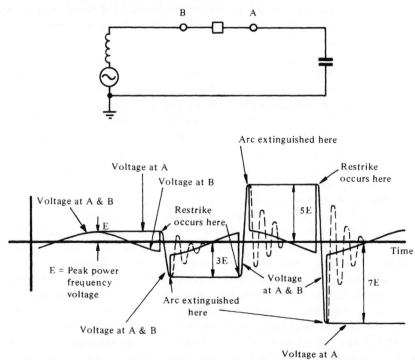

Fig. 11.1.2B *Hypothetical phenomena as a result of restrikes during the interruption of the capacity current of capacitors, unloaded cables or overhead lines*

Restrikes in circuit breakers do not occur in the regular fashion described above and, therefore, overvoltages do not reach the values shown. Slow circuit breakers with many restrikes may, however, produce undesirable overvoltages. It is significant with this type of overvoltage that the phenomenon develops while the circuit breaker is arcing after a restrike. The overvoltage thus affects plant on both sides of the circuit breaker and is not limited to the circuit switched out. A flashover or gap operation on either side of the circuit breaker puts a fault on the system. If the fault is on the line side, the circuit breaker will clear it (see Sections 11.4.3 and 11.5.1(iii)).

Up-to-date circuit breakers are in general free from restrikes, and therefore overvoltages of this kind do not represent serious problems on new installations. In addition to the two types of transient overvoltages discussed here there are other circuits configurations which can produce transient overvoltages as a result of switching operations and also of fault initiation and fault clearing.

When energising a line a voltage step is imposed on it. As explained in Sections 11.2.1 and 11.2.2, a travelling wave is initiated and at the open-circuited remote end of the line voltage doubling takes place. The magnitude of the initial voltage step depends on the instant of the circuit breaker closure and in the case of a 'dead' line cannot be more than the peak voltage (to earth) of the energising source. If, however, the line has 'trapped charge' this might increase the initial voltage step (e.g. in the case of highspeed autoreclosing on transmission lines). A similar situation might arise if a 'dead' transformer feeder is in a state of 'ferroresonance', i.e. it is energised through the capacitive coupling from a 'live' parallel circuit.[34,72] Energising overvoltages on the British distribution circuits and on existing transmission circuits are usually of little importance. However, energising transients may be of importance particularly on transmission systems of the highest voltages. On the 275 kV and 400 kV systems such energising overvoltages produced some flashovers on transformer feeders.[34,72] It is outside the scope of this Chapter to discuss in detail the subject of transient overvoltages caused by switching operations. Reference is made to some of the extensive literature on this subject containing further references.[29-36, 72]

(*b*) *Sustained internal overvoltages:* It is hardly practicable to design a large network in such a way that excessive lightning or switching surges do not occur under any circumstances. They have to be taken for granted and as will be seen later suitable surge protection has to be provided. On the other hand, sustained overvoltages as distinct from surges may cause very serious damage to expensive plant. Unfortunately, present techniques of overvoltage protection are not effective in such cases. It is, therefore, necessary to design the system so that large sustained overvoltages do not arise. A discussion of the system design aspects of this problem is outside the scope of this Chapter, particularly because resistance or solid earthing of the neutral as is the practice in Great Britain on most systems minimises the risk of sustained overvoltages. As there has been little trouble on this account in this country only brief reference is made to some typical conditions causing sustained overvoltages.

(i) *Neutral inversion:* The neutral point of a three-phase system under normal conditions is symmetrical in relation to the voltages of the three phases, that is the neutral is in the centre of the phasor diagram showing the phase voltages. Earthing of the neutral, therefore, ensures that the r.m.s. value of phase-to-earth voltages is practically the same for all three phases. The term 'neutral inversion' describes a condition when the neutral point assumes a position outside the perimeter of the triangle formed by the phase voltage vectors. If the neutral is earthed, the phase-to-earth voltages may greatly increase. The most common example of this condition is a badly designed star connected voltage transformer which has its starpoint earthed and which is connected to an otherwise unearthed highly capacitive circuit; a most unlikely condition on systems in this country. It is, therefore, not proposed to discuss the most intriguing phenomenon of neutral inversion here but reference is made to published literature.[37,38]

(ii)   *Arcing ground phenomena:* On systems without any neutral earthing, the current in a fault arc is determined by the capacitance of the system and the arc may be unstable. If the arc extinguishes and restrikes with a regular pattern, very high voltages may appear on the healthy phases. The mechanism of arcing ground phenomena has been the subject of a number of papers and text books but more recently some experiments cast considerable doubts as to whether the extinction and restrike of the fault arc can in practice follow the assumed pattern. As in this country, the Electricity Regulations do not permit the operation of a system without suitable earthing of the neutral the problem has little practical importance for us.

(iii)   *Resonance phenomena:* Under abnormal conditions resonance conditions may arise. Such abnormal conditions may arise as a result of the open-circuiting of one of two phases in a three-phase system, for example by faulty circuit breaker or broken conductors. A simplified diagram of an example is shown in Fig. 11.1.2C.

Excessive overvoltages can appear on two open-circuited phases of a system which is energised through one phase only of a faulty circuit breaker at a voltage V_1. Diagram (a) has been redrawn in two stages, (b) and (c), to illustrate more clearly the series resonance circuits. The circuit components producing resonance are shown on the right-hand side of the dotted line in diagrams (b) and (c), the latter being the equivalent circuit of (a). Resonance occurs when the inductive reactance $(\omega L + \omega L/2)$ equals the capacitive reactance $(1/2\omega C)$.

Under the oversimplified hypothetical conditions, the voltages on the open-circuited phases V_2 and V_3 could in theory be very high indeed, but in practice losses and saturation phenomena will limit their amplitude. Nevertheless, they may still be dangerously high.[39] The high inductance required for resonance may be obtained in a transformer or shunt reactor.

It should be noted that the phenomenon described requires that the star-point of the transformer (or shunt reactor) is not earthed, though the supply system neutral is solidly earthed in the example.

Another possible cause of voltage rise may occur when a large section of cable or a power-factor correction capacitor is disconnected from its normal supply system and is inadvertently energised through a very large inductive reactance, for example from the lower voltage side of a transformer which remains energised by some local source. Fig. 11.1.2D shows an example in which a generator is connected to the 11 kV busbars of a system which normally receives most of its supply through the

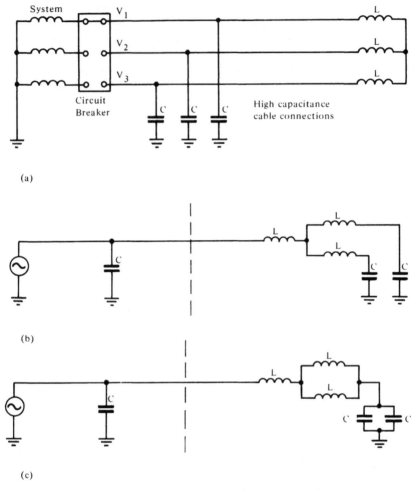

Fig. 11.1.2C *Simplified example of resonance conditions resulting from the open circuiting of two phases due to a faulty circuit breaker*

cabled 132 kV/11 kV transformer feeder. Excessive overvoltages may arise on the disconnected 132 kV side of the transformer and the cable if the circuit is inadvertently energised from the 11kV side. Diagram (*b*) shows the equivalent circuit of diagram (*a*) ignoring the transformer ratio. Series resonance occurs when the leakage reactance of the transformer is equal to the capacitive reactance of the cable.

(*c*)   *Temporary overvoltages:* This term has recently appeared in CIGRE and IEC documents.[40-42] It describes repetitive overvoltages maintained for several cycles. These overvoltages are of particular importance for ultra-high voltage systems, and stand between the transient and sustained overvoltages described in (*a*) and (*b*). They are in general of an oscillatory nature of relatively long duration, i.e.

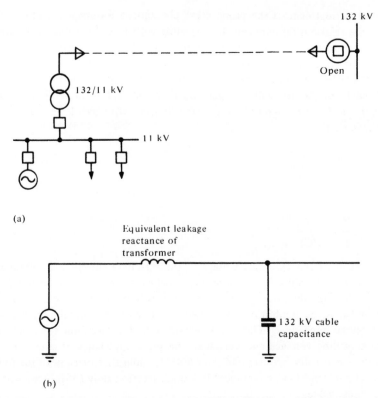

Fig. 11.1.2D *Resonance conditions on a cabled 132 kV/11kV transformer feeder when it is inadvertently energised from the low voltage side and disconnected from the 132 kV system*

hundreds or thousands of milliseconds. Temporary overvoltages usually originate from switching operations or faults, e.g. load rejection. Nonlinearity (saturation of transformers) and harmonic resonance amplifying the harmonics superimposed on the fundamental voltage is one of the typical types of temporary overvoltages.[60]

11.2   Travelling waves

11.2.1   Wave propagation along a transmission line without losses

When a voltage is applied suddenly to a transmission line (between two conductors or one conductor and earth), an energy wave will travel along the line with a speed approaching that of light. There will be a travelling electric field associated with the voltage on the line. As the line has capacitance, the voltage wave can advance only if accompanied by the current required to charge up the line. The voltage will result in an electric field and the current in a magnetic field. The propagating energy is equally divided between the electric and magnetic fields. While a precise mathematical treatment of the propagation phenomena is outside the scope of this Chapter, the relationship between the travelling voltage and current on a transmission line can be easily established by considering that the electrical energy on a length of line equals $CV^2/2$, where C is the capacitance of that length and V is the magnitude of the travelling voltage wave. On the other hand, the magnetic energy on that length of the line will be equal to $LI^2/2$, where L is the self-inductance of that length and I is the magnitude of the travelling current wave. It has already been stated above that

$$\frac{CV^2}{2} = \frac{LI^2}{2}$$

hence

$$\frac{V}{I} = \sqrt{\frac{L}{C}}$$

Thus the ratio of the travelling voltage to the travelling current is determined by the ratio of L to C. Since both are proportional to the length V/I does not depend on the length of the line. For convenience, $\sqrt{L/C}$ is termed the 'surge impedance' of the line and Z is the symbol normally used for it. With the formal introduction of the term 'surge impedance', Ohm's law remains valid for describing the relationship between travelling voltages and currents. The surge impedance of overhead transmission lines is usually between 300 and 500 Ω, while that of cables (due to their high capacitance and low inductance) is a small fraction only (of the order of one-tenth) of these values.

11.2.2   Reflections at the end of the line

When a travelling wave reaches the end of an open-circuited line, no current can flow out at the point of open-circuit, thus the magnetic energy must be zero there. The electric energy must, therefore, be doubled as a result of the voltage wave reflection or, in other words, at an open-circuit the original and reflected voltage waves have the same polarity and are superimposed on each other. As the current must be zero at the open-circuited end, the original and reflected currents are of opposite polarity and superimposed on each other.

If the line is short-circuited or earthed at the end, the role of the voltage and current is reversed. In this case, the voltage must be zero at the short-circuit and thus the electric energy is also zero. Consequently, all the energy must be in the magnetic field and the current doubled. Fig. 11.2.2A shows voltage and current reflection phenomena at open and short-circuited ends of a transmission line.

11.2.3   Discontinuities in surge impedance and junctions with infinitely long lines

When two transmission lines of different surge impedances are connected to each

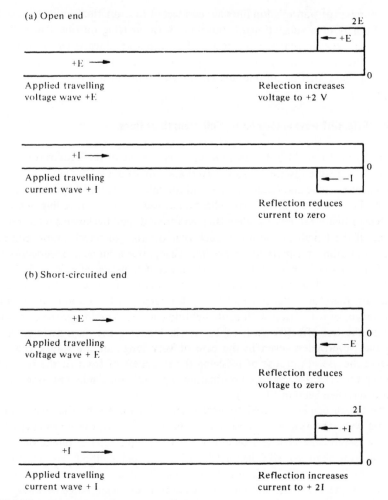

Fig. 11.2.2A *Travelling wave reflection phenomena at open and short-circuited ends of transmission line*

other, the travelling waves will partly pass to the second line and partly be reflected at the point of discontinuity. If the surge impedance of one line is Z_1, and that of the other Z_2, the voltage transmitted to the second line will be

$$E_2 = \frac{2Z_2}{Z_1 + Z_2} E_1 \qquad \text{(Ref.44)}$$

Thus if Z_1 is very high compared with Z_2, for example when an overhead line continues in a cable, the voltage wave is greatly reduced at the point where it enters the cable.

If a number of transmission lines are connected to a junction point (for example to the busbars of a substation), a travelling wave arriving on one line will see a reduction in surge impedance as a result of all the other lines being paralleled. If n overhead-line circuits of similar surge impedance are connected to a junction point, a travelling wave arriving on one line circuit with a magnitude of E will be reduced to $2E/n$.

11.2.4   Effect of waveshape and of finite length of lines

In Sections 11.2.1 to 11.2.3, travelling waves consisting of single steps (or so-called step functions) were discussed. Any waveshape can, of course, be approximated by a number of positive and negative steps and the reflections studied for each step. In practice, however, a further complication arises in that travelling waves are subjected to many reflections when they go forwards and backwards on a transmission line. If the length of the line is such that the time of travel is short compared with the duration of a particular travelling wave, after a number of reflections the line will be charged up in steps and the apparent benefits referred to in Section 11.2.3 gradually disappear. Relatively short lengths of cable connected to substations are therefore fully beneficial only for surges with durations in the microsecond range, as is the case with some lightning surges. The longer the duration of the surge the less effective such cables are in reducing the magnitude of voltage transients. Nevertheless, even in the case of very long-duration surges, the cables may serve the useful purpose of reducing the steepness of front of the surge, thus improving the effectiveness of co-ordinating gaps with their inherent long time lag to sparkover. (See Section 11.4.3.)

Using the principles referred to here briefly in an oversimplified way, for any travelling wave the response of various parts of a network can be calculated in a relatively easy but nevertheless tedious way. The simple rules of reflections permit graphical treatments by so-called lattice diagrams and methods are described in detail in the literature.[44,45] Unfortunately, the usefulness of accurate calculations is limited by the fact that lightning surges cover a very wide range of wave shapes and surge durations. Assumptions used in the calculations are, therefore, inevitably arbitrary. Results of such studies may nevertheless be helpful in assessing the statistical probability of the occurrence of overvoltages on certain parts of the system.

11.3   Insulation co-ordination

11.3.1   Fundamental principles of surge protection and insulation co-ordination

The external and internal overvoltages (often referred to simply as 'surges') discussed in the previous Sections may cause insulation breakdown on the various items of plant of the power supply system, causing usually a disturbance on the system or in worse cases serious damage to expensive equipment. Broadly speaking, any step to reduce the severity of such surge voltages could be regarded as 'surge protection'.

Protection against surge voltages must be related to the insulation strength of the plant affected or, in other words, the insulation has to be co-ordinated with the surge voltages which may arise. The main purpose of 'insulation co-ordination' is to ensure that the system characteristics and the protective devices applied are in such relation to the insulation of the various types of plant that, in service, reasonable freedom is obtained from supply interruptions and plant failures.[41,42]

11.3.2 Basic requirements

In order to avoid insulation failures, the insulation strength or 'insulation level, of the different types of equipment connected to the system has to be higher than the magnitude of transient overvoltages. As this magnitude is usually limited to a so-called 'protective level' by protective devices, it can be said in simple terms that the insulation level has to be above the protective level by a safe margin. Unfortunately, both the insulation strength and the protective level depend on a number of conditions and cannot normally be expressed in simple figures. The insulation strength of the various types of plant depends to a large extent on the waveshape, duration and repetition rate, and polarity of the overvoltages applied, while the protective level established by surge limiting devices, for example co-ordinating gaps and surge diverters (see Sections 11.4, 11.5 and 11.6), may depend not only on the waveshape and polarity but also on other factors, such as the magnitude of the surge current and the distance of the protective device from the plant to be protected.

11.3.3 Insulation and protective levels

The insulation strength or 'insulation level' is somewhat arbitrarily defined by overvoltage stresses which can be easily reproduced for test purposes such as the power-frequency and impulse overvoltage tests. Under specified conditions, 'protective levels' to which the magnitude of surges is limited by protective devices, such as co-ordinating gaps and surge diverters, are also established. It is necessary to bear in mind that service conditions may differ considerably from the assumptions upon which the insulation and the protective levels are established. Power-frequency and impulse overvoltage tests represent extreme conditions which do not normally occur in service. In catering for these extreme conditions, protection is provided against most types of surges occurring in practice but there are exceptions which should not be ignored entirely.

11.3.4 Relation between overvoltage tests and service conditions

Overvoltage tests originally consisted of the application of power-frequency voltages using large margins which experience proved satisfactory for various types of plant.

The power-frequency test voltages applied were often several times higher than the operating voltage. As power-frequency voltages on a properly designed and operated system could never approach such magnitudes, the only justification for these high power-frequency test voltages could be to prove the suitability of the insulation against transient overvoltages as discussed in Sections 11.1 and 11.2. The insulation strength of various items of plant may, however, greatly depend on the waveshape, duration and repetition rate of the voltage stress. The power-frequency voltage test to prove insulation strength against short-duration transients was, therefore, soon regarded as unrealistic and an impulse voltage test was introduced. Originally, the standard impulse voltage waveshape adopted in this country and in Europe was such that the magnitude of the voltage wave reaches its peak in 1 μs and in 50 μs thereafter it sinks to half value. This was called a 1/50 μs impulse wave and was adopted by European countries and also by the British Standards Institute. In the USA, 1·5/40 μs impulses were standardised and recently the International Electrotechnical Commission (IEC) laid down 1·2/50 μs waves but with such tolerance limits that both European and USA Standards comply with the specification.

The term impulse voltage has recently been extended considerably. Impulses with a front duration up to a few tens of microsecond are in general considered as lightning impulses and those having front durations of some tens up to thousands of microseconds as switching impulses according to IEC Publication 60-1. The 1·2/40 impulse is now called Standard Lightning Impulse. The Standard Switching Impulse is defined as 250/2500 impulse, but IEC 60-1 states that when such an impulse test is not considered sufficient or appropriate impulses of 100/2500 and 500/2500 are recommended. IEC Publication 99-1 Lightning Arresters, requires tests with 30 μs front time. Proposals have also been made to simulate oscillatory overvoltage transients by an oscillatory impulse.

The lightning impulse tests could be regarded as reasonable representation of severe lightning overvoltages but it is necessary to realise that lightning surges often have considerably longer wavefronts than the 1·2 μs, and in this respect they may come close to switching impulses. On the other hand, some types of switching overvoltages may have very steep fronts thus resembling in this respect the lightning impulse. The duration of lightning overvoltage reaching a substation might be extremely short, e.g. a few microseconds in the case of back-flashovers on the line, but at the other extreme the duration might extend to hundreds of milliseconds in the case of multiple strokes with no line flashovers. Switching overvoltage transients are often of an oscillatory nature and particularly for distribution plant they are often represented by power-frequency tests. Neither the lightning impulse, switching impulses or the power-frequency tests can simulate all types of insulation stresses which can occur on a distribution or transmission system.

The generation for test purposes of surges of the most onerous waveshape may present formidable practical difficulties, particularly at very high voltages. Available test plant is seldom equipped with the necessary components required to generate (e.g. oscillatory transients or very long) duration surges. Neither is the theoretical background fully clarified to decide on standard waveshapes for switching surge simulation applicable to all practical cases.

It has to be appreciated that standard test procedures must be based inevitably on somewhat arbitrary assumptions and cannot cover all possible service conditions. The impulse and power-frequency overvoltage tests together as recommended in the relevant BSI and IEC[41,42] standards represent, however, a reasonable safeguard that the various types of plant will withstand voltage stresses which can occur in service including switching surges.

11.3.5 Practical choice of insulation levels

Generally, it is not practicable to construct a power system on which outages would never take place due to insulation breakdown. A small number of flashovers in air have to be tolerated, particularly if permanent damage is not likely to be involved. On the other hand, the risk of insulation breakdown causing damage to expensive equipment has to be eliminated to as large an extent as is economically practicable.

The impulse insulation levels (verified by switching and lightning impulse overvoltage tests) are normally established at a value approximately 15-25% above the protection level to which the surges are limited by surge arresters or protective co-ordinating gaps.

It is much more difficult to settle the power-frequency insulation level since the only service condition with which power frequency overvoltage tests have some very distant relationship are switching overvoltages. It has to be noted here that, although for the higher transmission voltages switching impulse voltage levels are now specified at distribution voltages it is still usual to rely on power-frequency insulation levels to ensure that the insulation will withstand switching overvoltages.

The magnitudes of power-frequency test voltage have been determined largely on experience, and although (unlike the case of impulse tests) there may not be any clear scientific basis for these tests, they are nevertheless not quite so meaningless as has sometimes been suggested. They have proved invaluable in demonstrating the soundness of design and manufacture of plant and they provide safeguards against unknown stresses. Thus briefly they maintain good engineering practice as proved by service experience. It is to be expected that in the future more sophisticated testing techniques will be adopted to simulate more realistically various onerous overvoltage conditions occurring in service. Switching impulse overvoltage tests have already been introduced for rated voltages of 300 kV and above. It has to be borne in mind, however, that most overvoltages on a power system are of a composite nature consisting of power-frequency voltages with transients superimposed on them. To simulate these by composite tests may, however, prove to be too complex and unjustifiable in practice.

There have been recent trends, mainly abroad, to reduce both inpulse and power-frequency insulation levels on account of the availability at the present time of surge arresters with greatly improved protective charateristics. So far as the impulse insulation levels are concerned, there might be some justification for a reduction, since overvoltages approximating to the waveshape used in standard impulse tests can normally be limited by modern surge diverters. If, however, the power-frequency test voltages are also reduced, certain stresses which may occur in service (for example, phase-phase and temporary overvoltages) may not be covered. This aspect has to be considered carefully to ensure that any reduction is fully justified.

There is no doubt, however, that a reduction of the insulation levels would reduce safety margins, particularly in relation to internally generated transient overvoltages. Also in the remote case of surge currents in excess of the rated current of the surge diverter (see Section 11.6.2), the risk of breakdown is increased if a reduced insulation level is adopted. Nevertheless, such use of a reduced insulation level may be a practical proposition if careful attention is paid to stresses which are not adequately covered by the standard impulse and power-frequency tests. This usually requires detailed study of system conditions for certain items of plant and might involve special tests and a check that the circuit breakers employed are suitable from the point of view of not producing excessive overvoltages.

11.4   Protection against external overvoltages

Flashovers on overhead lines caused by lightning surges seldom cause permanent damage provided the line is equipped with high-speed protection so that the duration of the power-frequency fault current in the fault arc is limited.

In such cases, the main consequence of such a fault is an outage. A small number of such outages has to be tolerated on economically designed networks. On the other hand, lightning transients of excessive magnitude reaching a terminal station could cause breakdown of internal insulation of expensive plant, and such damage has to be avoided. Accordingly the steps taken to alleviate the effects of lightning surges on power systems can conveniently be discussed in two main groups.

(a)   Reducing the magnitude, the front steepness or the frequency of occurrence of surges at the point where they are initiated by providing shielding earth wires on overhead lines. Shielding of substation plant can also be discussed here.

(b)   Limiting the magnitude of surges at points where they could be most harmful. This requires protection of plant at substations, and whilst the inherent protective features of the system may be utilised, it is still usually necessary to limit the magnitude of surges at substations by limiting devices, for example protective gaps or surge diverters.

11.4.1   Shielding of overhead lines and substations

As seen in Section 11.1.1(b), the highest transient overvoltage will arise on an overhead line when a lightning stroke terminates on the phase conductors. To reduce such risks, overhead lines are often provided with a shielding earth-wire or wires above the phase conductors. It is often convenient to refer to the so-called 'shielding angle' of the earth-wire, that is the angle included between the vertical through the earth-wire and a line joining the earth-wire to the outermost line conductor. Fig. 11.4.1A illustrates the shielding angle. The lower the angle, the better is the shielding efficiency. Practical experience shows that with a shielding angle of 35-45°

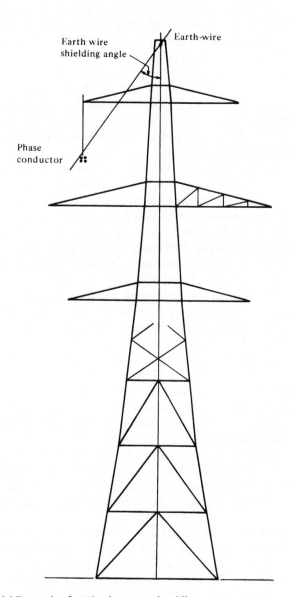

Fig. 11.4.1A *Shielding angle of earth-wire on overhead line*

most, but not all, lightning strokes are intercepted by the earth-wire.

In the past, the shielding efficiency was related simply to the shielding angle. More recent studies, however, have confirmed that the problem is much more complex and that the statistical probability of the phase conductors being struck depends not only on the shielding angle, but also on a number of other factors, among which are the distance of the shielding earth-wire from the phase conductors and the height of the line. Also, characteristics of the lightning stroke itself may have an influence on the shielding efficiency, for example the lower the current intensity in the lightning stroke the less effective is the shielding, since strokes with very low current intensity may approach the phase conductors almost horizontally. To go too deeply into this subject would be outside the scope of this Chapter, particularly because the problems are still far from being fully clarified.[2-26] Nevertheless, it can be said for practical purposes that with the line construction generally employed in this country, a shielding angle of 45° has proved to be satisfactory and 30° could be regarded as very effective shielding: 35° has been adopted for new 400 kV construction.

Reference was made in Section 11.1.1(*b*) to so-called back-flashover which can take place from the earth-wire or tower to the phase conductors when a tower or the earth-wire is struck by lightning. Shielding earth-wires cannot eliminate such back-flashovers. Full advantage of shielding by earth-wires for reducing lightning outages can, therefore, be taken only if the insulation of the overhead line is sufficiently high to prevent such back-flashovers. This is seldom practicable for lines operating at 33 kV or lower voltages as, in addition to back-flashovers, induced voltages from nearby strokes are also likely to produce flashovers.

Although the provision of earth-wires on such lines does not, therefore, materially reduce the number of lightning outages, there may nevertheless be other very good reasons for providing a continuous return conductor in the form of an earth-wire. The provision of an earth-wire may also reduce the severity of lightning transients reaching terminal equipment and this has to be taken into account.

A materially increased insulation of the overhead line, for example by utilising the extra insulation of wood poles, could reduce lightning outages on 11 kV and 33 kV lines but, in practice, the arrangements of stay wires, steel cross arms, etc. may not permit such potential advantages to be exploited since mechanical, constructional, operational and safety considerations may be of greater importance than that of maximum insulation strength against lightning transients.[25]

In respect of 11 kV and 33 kV lines, therefore, a judicious application of protection, autoreclosing, fusing policies, can provide more improvement in respect of supply interruption, or damage to equipment than increasing the insulation level.[25a,g,h,n]

On overhead lines for 132 kV and higher voltages, the insulation level is so high that back-flashovers seldom occur if the tower footing resistances are kept low. Advantage can, therefore, be taken of the shielding effect of earth-wires. Conditions could be regarded as acceptable if the tower footing resistance of most of the towers is less than say 30 Ω for 132 kV lines and less than, say, 80 Ω for 275 kV lines. No precise limit can be laid down since the risk of back-flashovers can be expressed only in terms of statistical probability. Even if tower footing resistances well below the figure quoted are obtained, the risk of back-flashover cannot be entirely excluded. On the other hand, tower footing resistances in excess of the figures quoted are likely to show their adverse effects on the lightning performance of overhead lines. In extreme cases of very high tower footing resistances, the advantages of shielding by earth-wire may be completely lost as far as lightning performance is concerned. Tower footing resistance may be reduced by providing extra earthing electrodes usually in the form of driven rods or buried counterpoise. High footing resistance in the case of a few towers may be acceptable in view of the statistical nature of the risk involved. Engineering judgement in assessing how exposed the towers are to lightning has to be exercised in such cases. In the case of a tower on the top of a hill or in the case of an exceptionally tall tower (for example river crossing tower) every practicable step has to be taken to keep the footing resistance low.

From the sometimes contradictory publications relating to the various features of the lightning stroke and the phenomena involved (direct strokes, back-flashovers, induced overvoltages etc.) it should be mentioned here that successful experiments have been carried out in France[43] and other countries, in which real lightning strokes were initiated at a testing site by firing into thunderclouds rockets with a trailer wire. The advantages of such controlled lightning strokes are that very elaborate, sophisticated instrumentation can be used to record the characteristics of the strokes and also to study the effects of direct strokes and induced surges etc. There may be some doubts about the severity of such artificially (and therefore prematurely) induced lightning activity. Nevertheless, the data to be obtained are likely to be most useful for clarifying disputed aspects of lightning performance of power lines.

The shielding of substations is a somewhat controversial matter. Direct lightning strokes to the phase conductors at substations may cause serious damage to the insulation of expensive substation equipment. For this reason, the provision of shielding earth-wires over substation plant is regarded in many countries as an absolute necessity. On the other hand, the area of substations is small in relation to that of the system as a whole and the probability of a lightning stroke terminating there is remote. Adjacent earthed structures may further reduce the risk of strokes direct to phase conductors. In this country, shielding wires in substations erected in the early years of the 132 kV Grid led to maintenance difficulties and were later abandoned. Subsequent service experience has been entirely satisfactory and has not indicated an undue risk of direct strokes to phase conductors. This may not be applicable, however, to countries with more onerous lightning conditions.

11.4.2 Surge protection by effective system layout

As discussed in Section 11.2, when a travelling wave reaches an open circuit on the line (a practical case is that of a transformer feeder) the voltage may be doubled. On the other hand, when a surge reaches a busbar to which a number of other circuits are connected its magnitude will be reduced since the surge impedances of the other circuits are effectively paralleled.[44,46] While the 'self protecting' feature afforded by having a number of circuits connected to the busbars is quite noticeable when these are overhead lines, it is even more so when they are cable circuits, as the surge impedance of cables is very small compared with that of overhead lines. The surge impedance of overhead lines is of the order of 400 Ω while the surge impedance of cables may be one tenth of this value.

From the above, it can be seen that travelling waves represent a greater danger at the transformer end of a transformer feeder circuit than at a major busbar station which is a junction point for several power lines. Accordingly, there is more need for efficient surge limiting devices at the transformer end of a transformer feeder circuit than at a major substation. For this reason, terminal plant of a transformer feeder should preferably be protected by a surge arrester while the much more primitive co-ordinating gap may be sufficient at major substations, particularly if some of the connections are cabled. It has to be emphasised here that the magnitude of travelling waves reaching the cable is reduced not during its travel through the cable as is sometimes mistakenly assumed. The reduction occurs at the junction point (that is at the busbar) due to the change of surge impedance. All the circuits (even those which are not cabled) connected to the junction point will thus benefit from the inherent surge protective effects of cables.

11.4.3   Voltage limiting devices

The simplest surge voltage limiting device is a rod gap connected between the earth and the terminal of the apparatus to be protected. Fig. 11.4.3A shows co-ordinating gaps mounted on 132 kV/275 kV autotransformers. The arcing horns shown on the 132 kV bushings serve the main purpose of protecting the metal fittings on both the live and earthy ends of the bushing by providing a suitable anchorage for the fault arc (which may be initiated by surface pollution of the bushing). Another function of the arcing horn is to divert the arc from the porcelain surface. The higher the rated voltage of the system (that is the longer the porcelain) the less efficient this arc diverting function becomes and for this reason the arcing horns have been omitted from the 275 kV bushings. If a rod gap is used as the main surge limiting device, thus co-ordinating the protective level, it is called a 'co-ordinating gap'. Surge voltages above a certain value (depending on the setting of the gap) will cause a flashover followed by the collapse of the surge. The advantages of the co-ordinating gap are its cheapness and simplicity. Its main disadvantage is that the operation produces an earth fault on the system resulting in an outage. Also the protective characteristics of a co-ordinating gap leave much to be desired. There is a large dispersion in flashover values aggravated by polarity effect, the flashover value being higher for surges of negative polarity. Due to the large time lag of the

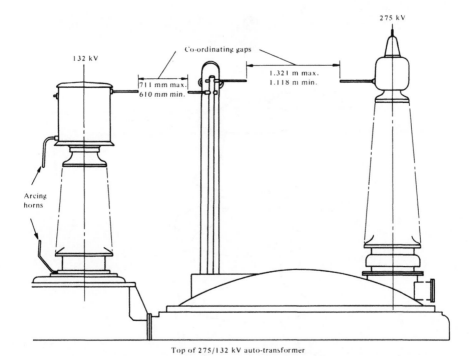

275 kV

132 kV

Co-ordinating gaps

711 mm max.
610 mm min.

1.321 m max.
1.118 m min.

Arcing
horns

Top of 275/132 kV auto-transformer

Fig. 11.4.3A *Typical 132 kV and 275 kV co-ordinating gaps mounted on a 275/132 kV autotransformer*

co-ordinating gap, breakdown might occur on solid or liquid insulation before the gap operates. At 33 kV and lower voltages, simple single co-ordinating gaps could easily be bridged by birds, vermin etc., and their use is, therefore, not recommended. This shortcoming is overcome by the use of duplex gaps, where the gap consists of two sections on opposite sides of an insulator. (Fig. 11.4.3B.)

A further development of the simple co-ordinating gap is the triggered gap. It has a reduced time lag and thus provides improved protection for internal (solid or liquid) insulation. A simple design of a triggered gap has been developed for the 11 kV system shown in Fig. 11.4.3C.[25(d)] The device containing the two outer main electrodes and the closely spaced inner auxiliary electrode pair represents essentially an unbalanced capacitive divider. If a sufficiently high impulse is applied to the terminals, the inner auxiliary electrodes spark over triggering breakdown across the main electrodes. A large number of such gaps have been installed on a trial basis by the electricity companies and have provided satisfactory service experience and more extended use of these devices is expected. Another possible application of triggered co-ordinating gaps has been proposed for metalclad substations using SF_6 insulation where the dispersion in both sparkover voltage and time lag is larger than in air. Such a triggered gap in SF_6 is described in a CIGRE Paper.[47]

The disadvantages of the co-ordinating gap are largely eliminated in surge

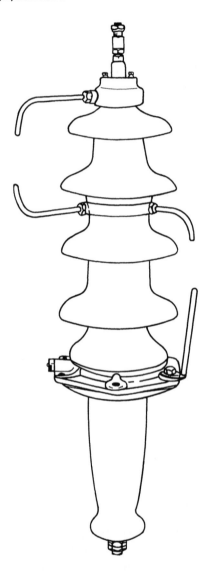

Fig. 11.4.3B   *33 kV bushing with duplex gap*

diverters, also called lightning arresters or, according to the latest IEC term, surge arresters, which consist of silicon carbide resistors having nonlinear voltage current characteristics and which are connected in series with multiple gap assemblies. Fig. 11.4.3D shows the principle of a surge arrester. When a surge voltage of excessive magnitude is superimposed on the power-frequency voltage, the gap assembly flashes over and the voltage across the arrester collapses to a value equal to the product of the surge current through the arrester and the resistance of the arrester.

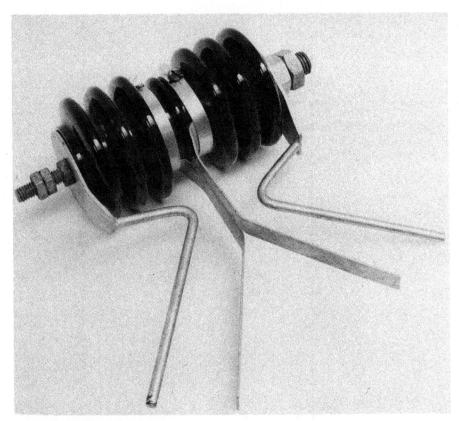

Fig. 11.4.3C *11 kV triggered gap*

Due to the nonlinearity, the resistance is very low for high surge currents. When the transient voltage disappears, a relatively small power-frequency follow current will flow (the resistance is very high at normal power-frequency voltages). The power follow current is extinguished at or near the first current zero.

The gaps in a surge arrester can be designed for high consistency of flashover values and for a small time lag in operation.

Although, on the one hand, it is desirable to reduce the voltage drop across the surge arrester during the flow of heavy impulse currents, on the other hand it is necessary that, at power-frequency voltage, the main nonlinear resistors should limit the follow current to a value which is low enough for it to be extinguished by the gaps. The power-frequency voltage at which the gaps are capable of extinguishing the power-frequency follow current is called the *rated voltage* of a surge arrester. The rated voltage of a surge arrester must not be less than the maximum possible power-frequency voltage across the arrester when it is called upon to operate, otherwise the power-frequency follow current will not be interrupted at the first current zero and the arrester will be destroyed. For any given design of surge arrester the rated voltage determines the number of gaps and series nonlinear

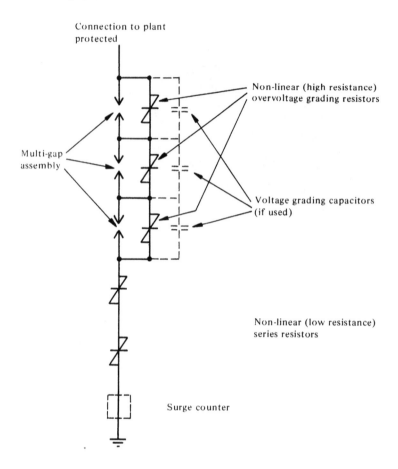

Fig. 11.4.3D   *Diagram of working principle of a surge arrester*

resistors and thus the protective level is directly proportional to the rated voltage of a surge arrester.

The requirements for improved protective characteristics and for increased reliability are to some extent contradictory. Lowering the flashover voltage of the gaps and the resistance of the series nonlinear resistors improves the protective characteristics but tends to make the interruption on the power-frequency follow current more difficult, thus reducing the reliability. Increasing the degree of non-linearity of the resistors is one way of solving this problem. Another way is to install gaps with artificially increased arc quenching properties, for example utilising the magnetic field resulting from the power-frequency current to propel the arc around a circular gap.

In the early designs of surge arresters the voltage drop in the gaps was insignificant, both during the passage of the surge current and also of the follow current. The magnitude of the follow current was determined by the nonlinear series resistors. Extinction of the follow current required such a number of nonlinear series resistors as to limit the follow current to a value at which after passing through current zero the gaps were capable of withstanding the returning voltage, i.e. the arrester resealed.

The efficiency of overvoltage limitation during heavy surge currents required however a low voltage drop across the arrester, i.e. as few nonlinear resistors in series as compatible with the requirement of limiting the follow current.

A practical solution of these contradicting requirements was the adoption of the so-called active gaps, also referred to as current limiting gaps. Thes gaps are designed on the principle that, after the passage of the high surge current, the length of the arc is extended, thus producing a material voltage drop during the follow current period, thus the power-frequency voltage across the series nonlinear resistors is reduced with the consequent reduction of the follow current. The various designs of such active gaps apply a magnetic field to the arc forcing it to extend in length.[48],[49]

Suitably designed active gaps can produce such high voltage drop that the arrester can reseal even against sustained direct voltage.[50]

It has to be noted that the energy dissipation in the active gaps can be material and the design should ensure that the life of these gaps is not adversely affected.

The use of active gaps permits a low ratio between the protective level (i.e. the level to which the overvoltage is limited) and the power frequency voltage at which the surge arrester must be capable of resealing. This is an important requirement at transmission voltages where the ratio of insulation level and system voltage is considerably lower than at the lower distribution voltages (see Sections 11.6.2 and 11.6.4 and Table 11.6.4A). For this reason, all up-to-date surge arresters used on transmission systems incorporate active gaps but not always at distribution voltages.

At sites where industrial or saline pollution and high humidity occur, as is the case frequently in the UK, the nonuniform heating affect of the leakage currents on the wet polluted surface of the porcelain housing of the surge arrester produces dry and wet bands. Local sparking occurs across the dry bands with persistent step changes of the voltage distribution on the polluted surface of the porcelain housing. The stray capacitance between the polluted layers on the surface of the porcelain housing and the internal gaps may cause the gaps to sparkover at normal system voltage. If this process is repeated the arrester might be overheated and finally fail.[51-53]

Tests to establish the performance of a surge arrester under polluted conditions require keeping it energised at a polluted site for a long period (including two winters, i.e. at least 18 months). Such natural pollution testing stations exist in various countries. However, natural pollution testing is a very lengthy process and a number of artificial pollution tests have been proposed and are considered by the IEC. It has been found that greasing of the housing eliminates in most practical cases the risk of pollution failure of the surge arrester.

Naturally, surge arrester designs are subject to continued improvements just as other technological products. The protective efficiency depends on the degree of nonlinearity of the resistor blocks. Using zinc oxide as the base material such nonlinear volt/ampere characteristics have been achieved that the series gaps can be omitted as the current at normal system voltage is insignificant (in the microamp region). Such gapless surge arrester installations have now been in operation satisfactorily for several years.[54,55]

Co-ordinating gaps and lightning arresters are most effective when they are installed in the immediate vicinity of the plant to be protected. It is, therefore, the present trend in extra high voltage transmission systems to install these devices at the transformers (see Fig. 11.4.3A), the most expensive and probably the most vulnerable items of plant. To cater for the case where the plant may be situated at some distance from the protective devices and also to obtain some safety margin, it is desirable to fix the lightning impulse insulation level at least 20-25% above the voltage value to which such surges are limited by the protective devices. For switching impulse voltage, the margin should be at least 15%.

Reference should be made here to a device whose performance lies between that of a co-ordinating gap and a surge arrester. This is the expulsion tube, which consists of an insulating tube, lined with a fibrous insulating material. Within the tube an air gap is arranged. The tube is installed in such a way that one end of it forms part of an external air gap. The protective level is determined by the flashover value of the external and internal air gap in series. When the tube operates, the current in the power arc produces an evolution of gas from the fibre. The blast of high pressure gas quenches the power follow current at first current zero. These tubes, which have been mainly used in the USA, are restricted to systems with a low fault level and are now mainly of interest for installations designed in the past.

11.5   Protection against internal overvoltages

11.5.1   Protection against switching transients

The magnitude of switching transients depends on the circuit parameters and the performance characteristics of the circuit breaker. In the case of switching the magnetising current of shunt reactors or unloaded transformers, the magnitude of the switching overvoltage depends on any tendency of the circuit breaker to chop the current before it reaches zero value. In the case of switching a capacitor bank or an unloaded line, excessive overvoltages can develop only if the breaker restrikes.

There are two basic methods of dealing with switching transients:

(*a*)   Reducing switching surges by adopting circuit parameters which do not permit the development of excessive overvoltages when switching. The simplest method is to employ a circuit breaker which keeps the magnitude of switching transients within acceptable values.

(*b*)   Limiting the magnitude of switching surges by voltage-limiting devices, that is co-ordinating gaps or surge arresters just as in the case of lightning transients.

(i)   *Protection by utilising circuit parameters:* As seen in Section 11.1.2(i), the chopping of the magnetising current of a shunt reactor or unloaded transformer produces an oscillatory transient with a peak value of approximately $V_o = i_{ch}\sqrt{L/C}$. For a given value of the current at the moment of chopping, the transient voltage can be reduced by increasing the capacitance of the circuit. Additional capacitance may be provided by the insertion of an overhead line (for example, in the case of a transformer feeder) or even by quite a short length of cable.

From the formula quoted above, a superficial conclusion is drawn in a number of papers which suggest that the more capacitance is added to the circuit the lower will be the switching overvoltages. This conclusion is based on the incorrect assumption that the maximum current which a given circuit breaker is capable of chopping is a fixed value. In fact the maximum instantaneous value of current which can be chopped is a function of the circuit capacitance and, as shown by Young, for a given installation the addition of a little capacitance might in fact increase the switching overvoltages.[28]

To eliminate switching surges reliably, it is necessary to add sufficient capacitance to the transformer circuit to ensure that even chopping the peak value of the magnetising current does not produce excessive transient overvoltages.

It has to be borne in mind here that for a short period after switching in, the transformer magnetising current may reach values far in excess of the steady state current, and the most onerous conditions often arise when a circuit is tripped immediately after switching in. At voltages of 132 kV and above the insertion of a relatively short length of cable (say a few hundred yards) between the transformer and the circuit breaker or the insertion of an overhead line of a few miles (transformer feeder) usually adds sufficient capacitance to the circuit to eliminate the risk of excessive overvoltages when interrupting the magnetising currents, whatever type of circuit breaker is used.

In the case of interrupting the charging current of a capacitor bank or of an unloaded feeder (overhead) the phenomenon described in Section 11.1.2(ii) assumed that the supply side is purely inductive. When switching either overhead lines or underground cables, such conditions could arise on transmission or distribution systems when only one line is connected to a supply point such as a generating station or a transforming station. Such conditions are seldom encountered in this country due to the complex interconnected transmission and distribution systems. High capacitance on the supply side reduces the magnitude of overvoltages.

In this country the number of feeders connected to a supply point is usually sufficient to ensure that the capacitance on the busbars, that is on the supply side, will be large enough to prevent the build-up of excessive overvoltages, irrespective of any restrikes in the circuit breakers. This may be the reason why in practice no excessive overvoltages have been experienced in this country on account of switching either underground cables or overhead lines on the system, although tests with specially arranged circuits have clearly indicated the possibility of such occurrences.

(ii)   *Resistance switching:* The magnitude of voltage transients caused by switching inductive or capacitive currents can be reduced by resistance switching. This method carries out the switching in two stages, that is two sets of interruptors connected in series operate consecutively. A resistor is connected across the first set of interruptors. Thus, when the first set of contacts opens, a resistor is inserted into the circuit.

In the case of switching an inductive current, the resistance should be relatively high since its main purpose is to damp the oscillations in the first stage and to make the circuit essentially resistive when the final interruption takes place.

In the case of switching capacitor banks or feeders with high capacitance, the resistance should be sufficiently low to dissipate a substantial part of the charge left on the switched circuit before the next voltage peak is reached on the supply side (after about half a cycle).

It is thus seen that the requirements for maximum suppression of switching surges in inductive and capacitive circuits, respectively, are contradictory. Furthermore, short-circuit conditions may require the adoption of resistance values different from either of the optimum values for reducing switching surges. Acceptable compromise values may be found or nonlinear resistors may be employed.

The circuit breaker designer has to choose a suitable practical value for the various conditions the breaker will encounter in service. It is, however, outside the scope of this Chapter to go deeper into this subject. It has to be emphasised that resistance switching in itself is not necessarily a panacea for all types of switching overvoltages.

(iii) *Protection by voltage limiting devices:* As switching surges seldom have steep wave-fronts, voltage limiting devices, that is rod gaps and lightning arresters, are even more effective than in the case of lightning surges discussed under Section 11.4.3.

Both co-ordinating gaps and some surge arresters are suitable for protection against switching overvoltages, discussed under Section 11.1.2(*a*). Nevertheless, some special problems may arise and attention is drawn to the following.

When the magnetising current of a transformer or shunt reactor is 'chopped' the surge develops after the transformer or reactor is disconnected from the system. The system, therefore, cannot be affected by these surges nor will the operation of the co-ordinating gap at the transformer or reactor put a fault on the system. There is, however, some risk that a high frequency oscillation on the transformer connections, caused by the breakdown of the co-ordinating gap, triggers off a restrike across the partly open contacts of the circuit breaker. Such an occurrence applies a short-circuit to the system for half a cycle or so until the breaker finally clears. Very high mechanical stresses can be produced in this way in the arc-control device of an oil circuit breaker of obsolete design and damage may occur. There would be no adverse effect on air-blast breakers. Due to the damping effect of their nonlinear resistors, surge arresters are unlikely to trigger off such restrike. Modern station-type, as distinct from line-type, surge arresters can dissipate the energy of this type of surge and they provide entirely satisfactory protection.

Another type of surge occurs when de-energising a line which is unloaded, that is, it is already open at one end, the surge developing after repeated restrikes in the circuit breaker, that is, during periods when the circuit breaker contacts are arcing. Protection can be provided by co-ordinating gaps installed on either the station side or the line side of the circuit breaker, although in the latter case, a flashover of the gap may suddenly subject the partly open circuit breaker to full short-circuit current, and again damage may occur to oil circuit breakers of obsolete design. The usual British practice is to install co-ordinating gaps on the terminals of transformers, an arrangement which has the advantage of effectively placing the gap on the station side of any line being de-energised, but which has the disadvantage that flashover of the gap during de-energising of the line produces an earth fault on the transformer circuit, causing it to trip. In order to avoid such tripping, surge arresters can be used. It is important to note, however, that the thermal capacity of the silicon carbide discs of a surge arrester may not be sufficient to dissipate the energy of surges produced by the switching of very long, very high voltage overhead lines or underground cables. The result may be a failure of the arrester. This risk may be reduced by installing the surge arrester on the station side of the circuit breaker (for example at a transformer in the station), although in this position the protective efficiency will suffer slightly.

It should be mentioned here that resistance switching, apart from reducing the magnitude of overvoltages, has the added advantage of eliminating the risk of damage to oil circuit breakers referred to in the previous paragraphs.

11.5.2 Protection against sustained internal overvoltages

Sustained overvoltages, even though they do not reach the magnitudes associated with lightning and switching surges, can be most harmful, not only because they are sustained but also because they may be applied repeatedly. Co-ordinating gaps of usual settings cannot provide any protection, as they are unlikely to operate at the voltages concerned. Surge arresters may afford some small degree of protection although they would not survive the repetitive operations with sustained overvoltages. Thus there is no effective overvoltage protection against sustained overvoltages and it is necessary to design the system in such a way that no such overvoltages occur. Earthing of the system neutral directly or via a suitable resistor seems to be the safest method of avoiding sustained overvoltages.

11.5.3 Protection against internal temporary overvoltages

Temporary overvoltages could be regarded as being between the transient and sustained overvoltages. Their effect depends very much on their duration and so does the practicability of protection against them. If they last for a few cycles only, then surge arresters may provide suitable protection. A crucial point, however, is that they should be capable of dissipating the energy involved. This is a design requirement which is not catered for by existing BSI and IEC specifications.[56,57]

Temporary overvoltages of excessive magnitude and duration (i.e. seconds or more) present problems similar to those of sustained overvoltages. Fortunately on the British Systems it was found that their magnitude did not exceed values covered by power frequency tests, though this may not be the case for future UHV systems.

Particular attention should be drawn however to the effect of the temporary overvoltages on the choice of rated voltage and reseal requirements of surge arresters as discussed in Section 11.6.1.

11.6   Practical aspects and some special problems of insulation co-ordination and surge protection

11.6.1   Effect of system neutral earthing on insulation requirements

The method of neutral earthing will affect under earth-fault conditions the voltage rise on the healthy phase.[37] When the neutral is earthed by a resistor or a Petersen coil, the phase-to-earth voltage on the healthy phases can be equal to the phase-to-phase voltage. Even higher sustained overvoltages may appear under abnormal conditions on systems with unsuitable earthing of the neutral. On the other hand on so-called effectively earthed systems, for example when the neutral points of all transformers are solidly earthed, the maximum voltage to earth on the healthy phases cannot, during an earth fault, exceed 80% of the phase-to-phase voltage or 1·4 times the phase-to-earth voltage (this is, in fact, the definition of an 'effectively earthed system'). The maximum voltage to earth of the healthy phases during an earth fault may have to be taken into account in specifying the power-frequency insulation requirements for the plant used on such systems.

Nevertheless, the system neutral earthing has little direct effect on the magnitude of lightning surges and its effect is somewhat indefinite in the case of switching surges. At first sight, therefore, the impulse insulation level does not appear to be directly connected with the method of earthing the system neutral. There is, however, an indirect relationship when the system is protected by surge arresters, as will be seen from the following explanation.

One of the most important features of a surge arrester is its rated voltage, that is the power-frequency voltage at which the gaps are capable of extinguishing the power follow current, the magnitude of which depends on the number of nonlinear series resistors. For a given design of surge arrester, the 'rated voltage' is thus proportional to the number of gaps and to the number of gaps and series nonlinear resistors. On the other hand, the number of gaps determines the impulse voltage at which the arrester operates the number of nonlinear resistors the voltage drop across the arrester. Thus the protective level (expressed as an impulse voltage) of an arrester of a given design is proportional to the rated voltage.

A system can be regarded as effectively earthed if R_0/X_1 is less than 1 and X_0/X_1 is less than 3, where R_0 is the zero phase sequence resistance, X_0 is the zero phase sequence reactance and X_1 is the positive phase sequence reactance of the system.

For establishing the permissible minimum rated voltage of a surge arrester for a given system, an earth fault is usually assumed on one phase while the arrester operates on another phase as a result of a surge.[56,57] On a noneffectively earthed system under such fault conditions, the surge arrester may thus be subjected to the maximum phase-to-phase system voltage. On an effectively earthed system the maximum power-frequency voltage during an earth fault cannot reach more than 80% of the phase-to-phase voltage and, therefore, a lower rating can be adopted for the arrester. For this reason if a system is protected by surge arresters approximately 20% reduction is permissible on the basic impulse insulation level for effectively earthed systems. If the protection is by co-ordinating gaps, the above consideration is not strictly applicable. As, however, effective earthing does usually permit a reduction in the co-ordinating gap setting without introducing undue risks of flashovers, the same reduced basic impulse levels are used on systems protected by co-ordinating gaps.

The principle of assuming an earth fault on one phase of the system and a voltage surge together with a power-frequency voltage rise on the healthy phase for establishing the rated voltage of a surge arrester has been criticised on the basis that it assumes the unlikely coincidence of a number of contingencies and may lead to unnecessarily high rated voltages and thus insulation levels at the highest transmission voltages particularly in the ultra-high voltage ranges. However, present knowledge indicates that temporary overvoltages (see Fig. 11.1.2C) may follow a disturbance which causes surge arrester operation. The arrester thus may have to reseal against such temporary overvoltages. Although the reasoning for adopting a rated voltage corresponding to the increased voltage on the healthy phase during an earth fault may not have been entirely realistic, in practice it provided the right value. Studies on the special requirements, resulting from temporary overvoltages (e.g. resealing against temporary overvoltages and repetitive operations) indicates the need for updating the performance requirements and the tests to prove compliance with such requirements.[49,58,59] Temporary overvoltages due to transformer saturation and harmonic resonance on the a.c. system and surge arrester reseal requirements have received attention in recent years.[60] Revisions of the existing surge arrester specifications are in hand and substantial changes may emerge particularly for rated voltages higher than 245 kV. However, until firm recommendations emerge the existing standards[56,57] provide acceptable guidance in the choice of surge arresters on the existing British distribution and transmission systems.

The neutrals of the 132 kV, 275 kV and 400 kV transmission systems in Great Britain are solidly earthed at each transformer. These systems comply with the requirements of effective earthing and permit the adoption of 20% lower basic impulse levels compared with non-solidly earthed systems. Distribution systems using resistance or Petersen coil earthing are non-effectively earthed. Compliance with the requirements for effective earthing on solidly earthed 11 kV systems depends to a large extent on the resistance of the earth system connected to the neutral. As it is often impracticable to achieve sufficiently low resistance values such systems are generally treated as non-effectively earthed.

11.6.2   Choice of surge arresters and derivation of basic impulse insulation levels

As discussed before, the rated voltage of the surge arrester (diverter) must not be lower than the maximum power-frequency voltage which can appear on the healthy phases during an earth fault.

The rated current of an arrester is the maximum impulse current at which the peak discharge residual voltage (that is the voltage drop across the diverter) is determined. Thus as soon as the gaps operate, the voltage across the arrester is limited to the discharge residual voltage. The protective level is thus related to this voltage. Nevertheless, the voltage limiting action cannot commence until the gaps spark-over and, therefore, the protective level cannot be below the impulse spark-over voltage of the gaps. Furthermore, in the case of very steep-fronted surges, due to the time lag, the surge voltage may rise to values higher than the 1·2 50 μs spark-over voltage before the gaps operate, and in such cases the protective level cannot be below the wavefront impulse spark-over voltage, that is the protective level of a surge arrester is determined by

(*a*)   peak discharge residual voltage
(*b*)   maximum impulse spark-over voltage
(*c*)   maximum wave-front impulse spark-over voltage.

Whichever is the highest of these three quantities gives the protective level of the surge arrester. If very steep-fronted surges can be excluded by suitable system design, for example by very effective shielding of the live conductors of substations and overhead lines (at least for the last mile or so) against direct lightning strokes, (*c*) may be ignored. In good designs of surge arresters the aim is to keep the values of (*a*), (*b*) and (*c*) reasonably close to each other. It should be noted that the lightning and switching impulse protective levels may differ slightly.

The higher the surge current that can pass through an arrester for a given voltage drop (that is for a given peak discharge residual voltage) the better the overvoltage protection afforded. It is, therefore, customary to choose a rated current of 10 kA for arresters protecting expensive plant at the highest voltages and also at every major substation. Such arresters are often referred to as 'station type arresters'. To protect less expensive plant at lower voltages cheaper, arresters rated at 5 kA, 2·5 kA or 1·5 kA may be used.

To cater for the distance of some plant from the protective devices and also to obtain some safety margin, it is desirable to fix the basic impulse insulation level 20-25% above the voltage values to which the surges are limited by protective devices. The basic insulation level always applies to insulation strength between phase and earth. The insulation strength between phases or across the gaps of isolating devices should be higher than the phase-to-earth level. It might be desirable

Table 11.6.2A: *Requirements for surge arresters applied to systems in Great Britain*

1 Nominal system voltage kV (r.m.s.)	2 Maximum system voltage kV (r.m.s.)	3 Method of earthing	4 Max. voltage to earth of healthy phase during earth faults kV (r.m.s.)	5 Min. rated voltage of surge diverter kV (r.m.s.)
11	12·1	Non- effective	12·1	
22	24·2	"	24·2	
33	36·3	"	36·3	The nearest available
66	72·6	"	73	diverter rating above
132	145	Effective	145 x 0·8 = 116	those shown in col. 4
275	300	"	300 x 0·8 = 240	
400	420	"	420 x 0·8 = 336	

to grade the insulation levels of external and internal insulation for certain types of high-voltage plant. Table 12.6.2A lists examples of surge arrester characteristics based on British Standard 2914 surge diverters, as applicable to distribution and transmission systems.

Present designs of surge arrester possess better protective characteristics than the minimum requirements specified in BS 2914[57] and they may permit the adoption of insulation levels lower than those generally used.

However, it is important to bear in mind that generous insulation levels and surge arrester rated voltages as recommended by IEC, BSI Standards and those of the Electricity Supply Industry usually include a margin for ignorance, i.e. for overvoltages resulting from conditions or from phenomena not visualised or understood by the designers. A material reduction of protective and insulation levels, i.e. the elimination of this margin will therefore necessitate a close and careful analysis of system conditions and possibly the introduction of special tests. Sections 11.4.3, 11.5.3 and 11.6.1 refer to some aspects of the problems, particularly those relating to temporary overvoltages and consecutive operation of surge arresters. A detailed treatment of these problems is outside the scope of this discussion and in any case some controversial aspects still need further clarification.[72]

11.6.3 Clearances to earth between phases and across isolating gaps

It is impracticable to carry out impulse tests on plant after it is assembled on site.

The impulse strength of the clearances depends on the electrode configuration and minimum clearances can be established by trying to determine the worst configurations met in practice. There is some degree of uncertainty in the minimum figures which are put forward in the various recommendations and the recommendations themselves have undergone considerable changes during the last few years. In establishing minimum clearances, two schools of thought exist. As external flashovers across clearances are preferable to internal puncture of insulation, it has been suggested in some countries that clearances should be arranged so that flashover occurs in about 50% of the cases when impulses equal to the basic insulation level are applied. In other countries, including the United Kingdom, there is a tendency to specify withstand clearances.

The voltages between phases may be approximately 15-25% higher than those to earth because of the presence of power-frequency voltages. This percentage depends on how many times higher the basic impulse level is than the power-frequency voltage. Considerably higher percentage may arise when overvoltages of opposite polarity are present on two phases of the system. Recent studies suggested that in some cases switching overvoltages between phases could be more than 50% higher than to earth.[72, 73] The problem becomes most apparent at very high voltages where there is a tendency to keep the basic insulation level as low as is consistent with reliability. Higher than normal voltages can occur across isolating gaps if a surge occurs when the systems on the two sides of an isolating gap are out of synchronism. It is sound practice, therefore, to have higher impulse strengths and clearances in these cases, although for practical reasons such impulse levels have not been specified so far. Drafts are being considered by IEC and expected to be published in the near future. To verify phase to phase insulation levels, the use of two synchronised impulse generators is proposed, one delivering positive, the other negative impulses.

11.6.4 Standard insulation levels, clearances with recommended co-ordinating gap settings, or surge arrester ratings, or both

As an example of the application of the principles discussed in this chapter, some essential data relating to 11-400 kV systems of the distribution companies and the national grid are given in Table 11.6.4A.

It is emphasised that the Table is based on data to be found on the largest part of existing installations and not necessarily on new construction. It is important to note that British Standards Institute (BSI), International Electrotechnical Commission (IEC) and Electricity Supply Industry (ESI) Standards are continuously revised and in the case of practical design problems the latest publications should be consulted. Reference to these publications indicates that insulation levels employed by the Electricity Supply Industry may in some cases be higher than stipulated in the publications. It is also important to note that the basic data above relate insulation of substation equipment with no particular attempt to incorporate

Table 11.6.4A: *Typical examples of insulation levels, clearances and co-ordinating gap settings employed at substations in Great Britain*

Nominal system voltage kV r.m.s.	Lightning impulse withstand voltage kV peak	Switching impulse withstand voltage kV peak	Power-frequency withstand test voltage (1 min dry) kV r.m.s.	Minimum clearance in air to earth in	Minimum clearance in air between phases in	Co-ordinating gap settings in
11	100	—	29	8	10	2 x 1*
33	195	—	76	14	17	2 x 3¾*
66	350	—	150	27	31	15
132	550	—	300	44	50	26
	640	—	300	50	58	26
275	1050	850	485	84	98	46-52
400	1425	1050	675	120	140	60-70

*Duplex gaps. Simple rod gaps not recommended.

specific requirements for individual items. Attention is drawn here to an IEE paper which surveys the various types of overvoltages on the UK electricity system and discusses the philosophy of overvoltage protection and insulation co-ordination in light of service experience. It gives details of the relevant test requirements.[72]

11.6.5 Effect of rain, humidity and atmospheric pollution

The insulation strength of air clearances of the order shown above is not affected materially by rain, humidity, or industrial and salt pollution. Heavy contamination of insulator surfaces, however, particularly at times of high humidity, might cause flashovers at normal system voltage. In this country on the 132 kV transmission system, the number of outages on this account is of the same order as that due to lightning. The performance of an insulator can be improved by increasing the creepage distance of the surface of the insulator. In the case of so-called 'antifog' insulation, at least one inch creepage distance is required for each kV of the system phase-to-phase voltage. At least one-half of the creepage length should be 'protected', that is that portion of the insulator surface lying in shadow when the insulator is illuminated from a distance at 90° to its axis.

Under exceptionally heavy pollution conditions even this increased creepage distance may not be sufficient to eliminate completely the risk of pollution flashover. Greasing of the insulators, particularly at 132 kV and above, may be the only practicable method of achieving further reduction of this risk.

Since such flashovers are a consequence of the formation of continuous conducting surface layers formed of moisture and pollution products, it is clear that, if the insulator could be made water-repellent and the formation of a continuous film thereby prevented, no conduction processes could take place and flashovers should be eliminated.

Either hydrocarbon grease or silicone grease may be used as water repellents. The effective life of the hydrocarbon grease is limited and it has to be completely removed periodically and replaced with new grease. The interval between such replacements will vary with the degree of pollution but, generally, hydrocarbon grease can remain in service for about three years.

Silicone greases, although appreciably more expensive than their hydrocarbon counterparts, have the advantage of maintaining stability at considerably higher temperatures such as may be experienced on transformer bushings. The effect of applying the compound is two-fold. First, the silicone material will tend to migrate from the compound and enclose each pollution particle, thus preventing the formation of a continuous conducting path. Secondly, it will cause water which condenses or falls on the contaminated surface to separate into individual droplets, thus preventing the formation of a continuous moisture film. The effective life of the coating will depend on the amount and nature of the pollution, the rainfall and the scouring action of wind-borne sand or dust; and the required frequency of silicone regreasing can, therefore, be determined only by experience.

11.7   Probabilistic or statistical approach in insulation co-ordination

11.7.1   Statistical aspects of overvoltages and insulation strength

As seen in Section 11.3 and elaborated in Section 11.4, 11.5 and 11.6, the basic requirement of effective insulation co-ordination is that the insulation strength, i.e. the minimum voltage which the insulation will definitely withstand, should be higher, by a suitable margin, than the maximum value of overvoltage. The meaning of maximum and minimum in this context referred to by the IEC as the conventional method is somewhat indefinite since both are subject to random variations and they follow (at least in some respect) a statistical distribution.

The insulation strength is established by a limited number of tests and the fact that insulation withstands say 10-15 shots cannot prove that a larger number of test shots would not cause breakdown. The probability distribution of the breakdown voltage of external (air) insulation or according to the IEC term 'self-restoring' insulation can be established for any electrode configuration with great accuracy by a large number of systematic laboratory tests. Considerable difficulties arise however for internal or according to IEC terminology non-self-restoring insulation. The problem is partly that a test piece is destroyed by every breakdown making the exercise costly, partly that there are various factors in the manufacturing process which do not follow simple statistical rules. The new test pieces to which the subsequent test shots are applied cannot therefore be considered identical to those which were destroyed in the earlier tests.

Similarly, the occurrence of overvoltages of a certain magnitude whether external or internal can best be presented on a statistical probability basis. The probability distribution of lightning surges depends on the lightning severity of the area concerned, design parameters of the power line etc., and a number of arbitrary assumptions have to be used in any assessment. The probability distribution of internal overvoltages can be studied by computers, automatic recordings on the system, staged tests and again a number of arbitrary assumptions have to be made. Fig. 11.7.1A shows a histogram of energising overvoltages recorded on the receiving end of an approximately 75 mile long 275 kV line of the CEGB during a series of

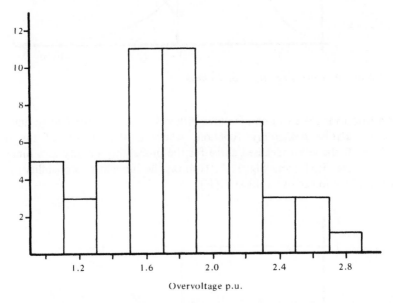

Overvoltage p.u.

Fig. 11.7.1A   *Histogram of overvoltages recorded on the receiving end of an approximately 75-mile long 275 kV line of the CEGB during a series of tests*

tests. An overvoltage factor of 2·8 p.u. was reached on one record only out of over 50. On 11 shots, 1·6 p.u. was reached and on another 11, 1·8 p.u. reached. For substation equipment the probability distribution of overvoltages is modified by the operation of voltage limiting gaps or surge arresters.

It is clear from the above that to establish with a limited number of test shots a 'withstand' insulation strength, i.e. a voltage level below which the insulation will never break down or to establish a maximum overvoltage level or protective level which will never be exceeded is not a practical proposition. It would be more meaningful to assess the 'risk of failure' of insulation by taking into account the statistical distribution of the two main factors the overvoltage distribution and the breakdown voltage distribution referred to by the IEC as the probabilistic approach.

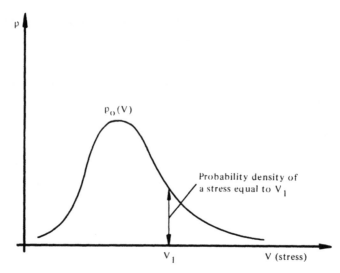

Fig. 11.7.1B   *Probability density of overvoltages*

Statistical variables can be presented either by a *probability density* function or by a *cumulative probability* function, which is the integral of the former. Fig. 11.7.1B shows an idealised curve for the probability density as a function of the overvoltage $P_0(V)$ and Fig. 11.7.1C shows the cumulative probability of break-down as a function of overvoltage $P_d(V)$.

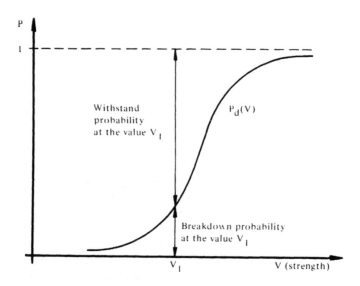

Fig. 11.7.1C   *Cumulative probability of breakdown of insulation*

11.7.2   Application of statistical distribution to insulation co-ordination

When the probability distribution is known for both the overvoltages and also for the breakdown of insulation the aggregate probability of a breakdown of insulation caused by the surges can be expressed in terms of statistical probability and the purpose of insulation co-ordination is to provide insulation for which the probability of breakdowns remains below a small figure, considered satisfactory in the particular case.

An oversimplified example of the elementary approach is shown here. In Fig. 11.7.2A the probability density of overvoltages $p_o(V)$ is shown together with the cumulative probability $P_d(V)$ of the breakdown of insulation. The probability of an overvoltage occurring within a small band of ΔV is $p_o(V)\Delta V$. The probability of a breakdown of the insulation at V is $P_d(V)$. The product of the two represents the risk of a breakdown for a band of overvoltages ΔV.

$$p_o(V) \cdot P_d(V) \cdot \Delta V$$

The total risk R of breakdown for the whole range of V is therefore:

$$R = \int_o^\infty p_o(V) \cdot P_d(V) \cdot dV$$

RISK EVALUATION

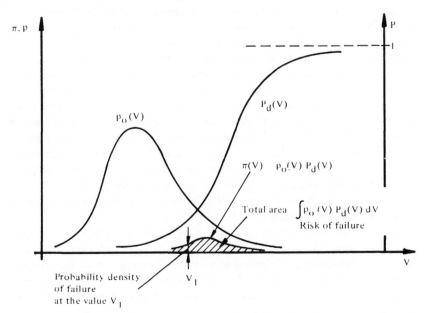

Fig. 11.7.2A   *Statistical determination of risk of failure for overvoltages*

The above example demonstrates the basic principle of the statistical approach to insulation co-ordination.[41],[42] Work is still in its early stages. Actual problems on a transmission system involve a very large number of variables which are being investigated and are outside the scope of these notes. The philosophy of statistical approach to insulation design should however appeal to research workers as it covers many unsolved problems and great potentials for analytical studies.[64-71]

If the results are used with common sense and engineering judgement such studies provide the most useful guide to the designer pinpointing problems requiring special attention.

11.8   Economic aspects

Generally, it is not practicable to construct a power system on which outages would never take place due to insulation breakdown. A small number of flashovers in air have to be tolerated, particularly if permanent damage is not likely to be involved. On the other hand, the risk of insulation breakdown causing damage to expensive equipment has to be eliminated to as large an extent as is economically practicable. The statistical methods discussed in Section 11.7 should be particularly suitable for assessing the relationship between failure rates and costs of plant.

Good engineering design requires the best use of available resources. Economic and engineering requirements cannot be separated from each other. The problems of reconciling costs and operational reliability have already been raised in Section 11.3.5 discussing problems of reduced insulation levels. The different methods of dealing with the adverse effects of surges can be combined and virtually any desired degree of freedom from outages and damage to equipment can be achieved depending on technical and operational requirements.

On the economic side, one has to consider the financial losses caused by supply interruptions and damage to plant which might result if in order to save capital expenditure inferior surge protective schemes, and/or unduly low insulation levels are used. Between the contradicting financial and technical requirements a reasonable compromise has to be found. Any investments in order to improve the performance of the system could be compared with the premiums of an insurance scheme. Since statistical methods are generally used by insurance companies for assessing risks, the combination of the statistical approach for insulation design can be combined, in the future, with statistical insurance methods. No doubt many details have to be analysed before such calculations will be accepted since the variables involved are very complex.

In assessing the consequences of losing a circuit temporarily or for a longer period of time due regard has to be paid to the general layout of the system, available spare capacities and alternative supplies. Tripping out of one circuit is usually of little consequence when alternative circuits maintain the continuity of supplies, but if this is not the case serious disturbances may result. Stability problems have to be considered.

Practical power systems are normally the results of long developments and the problem is very seldom that of producing a completely new scheme fully exploiting the possibilities of modern techniques. Existing plants during their useful life cannot normally be discarded without serious financial losses and the practical problem which the designer has to face in most cases during the development of power systems is to make the most economic and technically satisfactory use of plant and equipment already there.

11.9 Bibliography

1   B Schonland: 'Lightning and the long electric spark' (Presidential Address to British Association 'The advancement of science' 1962-1963, xix)

2   R H Golde: 'Lightning currents and potentials on overhead transmission lines' *J. IEE*, 1946, **93**, Pt.II, pp. 599-569

3   J H Gridley: 'The shielding of overhead lines against lightning' *Proc. IEE*, Part A, 1960, **107**, p.325

4   C F Wagner: 'The lightning stroke as related to transmission-line performance' (*Electr. Eng.*, May and June 1963, p.339)

5   K Berger: 'Elektrische Anforderungen an Hochstspannungs-Leitungen' (Bulletin de l'Association Suisse des Electriciens, 1963, **54**, p.749)

6   R Davis: 'Lightning flashovers on the British Grid' (*Proc. IEE,* 1963, **110**, pp.969-974)

7   R H Golde: 'Lightning performance of British high voltage distribution systems' (*ibid.*, 1966, **113**,(4), pp.601-610)

8   C F Wagner, 'The lightning stroke as related to transmission-line performance' Parts I and II (*Electr. Eng.*, May and June, 1963)

9   J M Clayton and F S Young: 'Estimating lightning performance of transmission lines' (*IEEE Trans.*, 1964, **PAS-83**, pp.1102-1110)

10 K Berger: 'Novel observations on lightning discharges: results of research on Mount San Salvatore' (*J. Franklin Inst.*, 1967, Special Issue on Lightning Research, **283**, pp.478-525)

11 M A Sargent and M Darveniza: 'Lightning performance of double circuit transmission lines' (*IEEE Trans.*, 1970, **PAS-89**, pp.913-925)

12 M A Sargent and M Darveniza: 'The calculation of double circuit outage rate of transmission lines' (*ibid.*, 1967, **PAS-86**, pp.665-678)

13 H R Armstrong and E R Whitehead: 'Field and analytical studies of transmission line shielding' (*ibid.*, 1968, **PAS-87**, pp.270-281)

14 G W Brown and E R Whitehead: 'Field and analytic studies of transmission line shielding, Part II' (*ibid.*, 1969, **PAS-88**, pp.617-226)

15 M A Sargent: 'The frequency distribution of current magnitude lightning strokes to tall structures' (PAS-91, No. 5, pp.2224-2229, 1972)

16   F Popolansky: 'Frequency distribution of amplitudes of lightning currents' (*Electra*, 1972, **22**, pp.139-147);

17   S A Prentice: 'CIGRE lightning flash counter' (*ibid.*, 1972, **22**, pp.149-169)

18   D W Gilman and E R Whitehead: 'The mechanism of lightning flashover on high-voltage and extra-high-voltage transmission lines' (*ibid.*, 1973, **27**, pp.65-96)

19   E R Whitehead: 'CIGRE survey of the lightning performance of extra-high-voltage transmission lines' (*ibid.*, 1974, **33**, pp.63-89)

20   M A Sargent: 'The frequency distribution of current magnitudes to tall structures' (*IEEE Trans.*, 1972, **PAS-91**, pp.2224-2229)

21   J R Currie, Liew ah Choy and D Darveniza: 'Monte Carlo determination of the frequency of lightning strokes and shielding failures on transmission lines' (*ibid.*, 1971, **PAS-90**, pp.2305-2312)

22   Liew Ah Choy and M Darveniza: 'A sensitivity analysis of lightning performance calculations for transmission lines' (*ibid.*, 1971, **PAS-90**, pp. 1443-1451)

23   M A Sargent and M Darveniza: 'Tower surge impedance' (*ibid.*, 1969, **PAS-88**, p.4754)

24   P Chowdhuri and E T B Gross: 'Voltage surges induced on overhead lines by lightning strokes' (*Proc. IEE.*, 1967, **114**,(12), pp.1899-1907)

25   IEE Conf. Publ. 108, (1974). 'Lightning and the distribution system' containing the following reports:

(a)   D R Aubrey: 'Co-ordination of fuses and circuit breakers during lightning storms'

(b)   W P Baker: 'Response of an 11 kV overhead network on induced overvoltages'

(c)   W P Baker; 'The impulse strength of 11 kV plant'

(d)   W P Baker and D F Oakeshott: 'Surge diverters and spark gap protection'

(e)   W Bowman and R N Ward: 'The duty capability of 11 kV auto-reclosing oil circuit breakers controlling overhead line feeders'

(f)   F Cornfield and M F Stringfellow: 'Calculation and measurement of lightning-induced overvoltages on overhead distribution lines'

(g)   J H Evans: 'An assessment of the lightning protection policies of the British distribution system'

(h)   J H Evans: 'Improving the lightning performance of the distribution system'

(i)   J H Gosden: 'Lightning and distribution systems – the nature of the problem'

(j)   J L Hughes and A K Hill: 'Field trials – their implementation and assessment'

(k)   D F Oakeshott: 'Lightning performance and protection practice of the British 132 kV system'

(l)   J J Seed: 'Fault current measurements on 11 kV overhead networks'

(m)   M F Stringfellow: 'Lightning incidence in the United Kingdom'

(n)   W G Watson: 'Selective application of pole-mounted reclosures and fuses to minimise consumer-interruptions'

(o)   W G Watson and H Lloyd: 'The relationship between reliability expenditure on 11 kV overhead circuits and the energy not supplied to consumer due to faults'

26   G W Brown: 'Joint frequency distribution of stroke current rates of rise and crest magnitude to transmission lines' (IEEE Paper F77 017-7, IEEE PES Winter Meeting, 1977)

27   K Berger and R Pichard: 'Die Berechnung der beim Abschalten leerlaufender Transformatoren, insbesondere mit Schnellschaltern entstehende Uberspannungen' (*Bulletin de l'Association Suisse des Electriciens,* 1944, **20**, p.551)

28   A F B Young: Some researches on current chopping in high-voltage circuit breakers (*Proc. IEE.,* 1953, **100**, p.337)

29   CIGRE Working Group 13-05. 'The calculation of switching surge I. Comparison of transient network analyser results' (*Electra,* 1971, **19**, pp.67-78)

30   CIGRE Working Group 13-05. 'The calculations of switching surge II. Network representation for energisation and re-energisation studies on lines fed by an inductive source' (*ibid.,* 1974, **32**, pp.17-42)

31   CIGRE Working Group 13-05. 'Switching overvoltages in EHV and UHV systems with special reference to closing and reclosing overvoltages of transmission lines' (*ibid.,* 1973, **30**, pp.70-122)

32   S Bernerd, C E Sölver, L Ahlgren and R Eriksson: 'Switching of shunt reactors: comparison between field and laboratory tests' (CIGRE Conf. Report 13-04, 1976)

33   G C Damstra: 'Influence of circuit parameters on current chopping and overvoltages in inductive m.v. circuits' (CIGRE Conf. Report 13-08, 1976)

34   L Csuros, K F Foreman and H Glavitsch: 'Energising overvoltages on transformer feeders' (*Electra,* 1971, **18**, pp.83-105)

35   J P Bickford and P S Doepel: 'Calculation of switching transients with particular reference to line energisation' (IEE Paper 5234P, Proc. IEE, April 1967)

36   A Clerici, G Santagostino, U Magagnoli and A Taschini: 'Overvoltages due to fault initiation and fault clearing and their influence on the design of UHV lines' (CIGRE Conf. Report 33-17, 1974)

37   J R Mortlock and C M Dobson: 'Neutral earthing of three-phase systems, with particular reference to large power stations' (*J. IEE,* 1947, **94**, Pt.II, pp.549-572)

38   B G Gates: 'Neutral inversion in power systems' (*ibid.,* 1936, **78**, pp.317-325)

39   C McNamara: 'Study of switching faults improves operation' (*Electr. World,* 26th February 1949, p.80)

40   R Gert, H Glavitsch, N N Tikhodeyer, S S Shur and B Thoren: 'Temporary overvoltages: their classification, magnitude, duration, shape and frequency

of occurrence' (CEGRE Conf. Report 33-12, 1972)

41   International Electrotechnical Commission Publication 71-1, 1976. 'Insulation co-ordination. Part 1: Terms, definitions, principles and rules'

42   International Electrotechnical Commission Publication 71-2, 1976. 'Insulation co-ordination. Part 2: Application Guide'

43   R P Fieux, C H Gary, A R Eybert-Berard, P L Hubert, A C Meesters, P H Perroud, J H Hamelin, J M Person: 'Research and artificially triggered lightning in France' (IEEE Paper F77 571-3, IEEE PES Summer Meeting, 1977)

44   L V Bewley: Travelling waves on transmission systems (John Wiley & Sons, New York)

45   E Q Boehne: 'Travelling wave protection problems' (Trans. AIEE, August 1954, p.920)

46   I W Gross, L B Le Vesconte and J K Dillard: 'Lightning protection in extra-high voltage stations' (*Electr. Eng.,* Nov. 1953, p.967)

47   B F Hampton, W J T Jinman, R J Meats, J J Fellerman and K F Foreman: 'A triggered co-ordinating gap for metalclad substations (CIGRE Conf. Report, 1978)

48   E C Sakeshaug: Current-limiting gap arrester – some fundamental considerations. (*IEEE Trans.,* 1971, **PAS-90**, pp.1563-1573)

49   A Schei and A Johansson: 'Temporary overvoltages and protective requirements for EHV and UHV arresters' (CIGRE Conf. Report 33-04, 1972)

50   G D Breuer, L Csuros, R W Flugum, J Käuferle, D Povh and A Schei: 'HVDC surge diverters and their application for overvoltage protection of HVDC schemes' (CIGRE Conf. Report 33-14, 1972)

51   L Torseke and T D Thorsteinsen: 'The influence of pollution on the characteristics of lightning arresters: theoretical aspects and artificial tests' (CIGRE Conf. Report 404, 1966)

52   E Nasser: 'The behaviour of lightning arresters under the influence of contamination' (*IEEE Trans.* Paper 31 CP 66-117, 1966)

53   A Schei and A Johansson: 'The effect of pollution and live-washing on surge arresters (CIGRE Conf. Report 33-02, 1974)

54   E C Sakshaug, J S Kresge and S A Miske: 'A new concept in station arrester design' (*IEEE Trans.,* 1977, **PAS-96**, pp.647-656)

55   M Kobayashi, M Mizuno, T Aizawa, M Hayashi and K Mitani: 'Development of zine-oxide non-linear resistors and their application to gapless surge arresters' (IEEE Summer Meeting, 1977)

56   International Electrotechnical Commission Publication 99-1. 'Lightning arresters, Part 1, Non-linear resistor type arrester for a.c. systems' 1970

57   British Standard Institute. BS 2914:1972. Specification for surge diverters for alternating current power circuits.

58   J D Phelps and R W Hugum: 'New concepts in the application of surge arresters for insulation co-ordination' (CIGRE Conf. Report 33-08, 1972)

59   E C Sakshaug, A Schei, A Clerici, G P Mazza, G Santagostino and A Taschini: 'Requirements on EHV and UHV Surge Arresters. Comparison of energy and

current duties between field and laboratory conditions by means of TNA simulation' (CIGRE Conf. Report 33-10, 1976)

60 L Csuros, Å Ekström and D Povh: 'Some aspects of HVDC insulation co-ordination' (CIGRE Conf. Report 33-10, 1978)

61 A Colombo, G Sartorio and A Taschini: 'Phase-to-phase air clearances in EHV substations as required by switching surges' (CIGRE Conf. Paper 33-11, 1972)

62 A C Legate, G E Stemler, K Reichert and N P Cuk: 'Limitation of phase-to-phase and phase-to-ground switching surges field tests in Bonneville power administration 550 kV system' (CIGRE Conf. Paper 33-06, 1976)

63 K H Weck, H Studinger, L Thione, A Pigini and G N Aleksandrov: 'Phase-to-phase and longitudinal insulation testing technique' (CIGRE Conf. Paper 33-09, 1976)

64 L O Barthold and L Paris: 'The probabilistic approach to insulation co-ordination' (Electra (CIGRE), May 1970, pp. 41-58)

65 G Carrara and L Marzio: 'Discharge probability under dielectric stresses' (Appendix V of Progress Report of Study Committee No. 8., CIGRE Conf. Report 33-01, 1968)

66 M Ouyang: 'New method for the assessment of switchings surge impulse insulation strength' (*Proc. IEE*, 1966, **113**,(11), pp.1835-1841)

67 G Carrara, E Occhini, L Paris and F Reggiani: 'Contribution to the study of insulation from the probabilistic point of view' (CIGRE Conf. Report 42-01, 1966)

68 L Paris, E Comellini and A Taschini: 'Insulation co-ordination of an EHV substation' (CIGRE Conf. Report 33-08, 1970)

69 V Vyskocil and J Fic: 'Results of a statistical investigation of overvoltages in the Czechoslovak systems' (CIGRE Conf. Paper 33-01, 1972)

70 C Dubanton and G Gervais: 'Switching overvoltages when closing unloaded lines: effect of power and system configuration, statistical distribution' (CIGRE Conf. Paper 33-05, 1972)

71 T Kawamura, T Kouno, H Mitani, S Kojima, K Harasawa, F Numajiri and H Ishihara: 'Statistical approach to the insulation co-ordination of substations against lightning overvoltage' (CIGRE Conf. Paper 33-06, 1974)

72 L Csuros and K F Foreman: 'Some pratical aspects of overvoltages on the CEGB transmission system' (*IEE Proc. E, Gen. Trans. Distrib.*, 1980, *127*, (4), pp.248-261)

73 K H Schneider: CIGRE Working Groups 33-02, 33-03, 33-06, Task Force 33-03-03: 'Phase-to-phase insulation co-ordination' (*Electra*, 1979, *64*, pp.137-236)

Index